Relativity For Beginners, The Special & The General Theory

The 3 Bizarre Discoveries You Must Know
To Master Einstein's Relativity Fast,
Revealed Step-By-Step (In Plain English)

By John Stoddard

**Praise for *Relativity for Beginners,
The Special and The General Theory***

"Vivid stories, completely captivating. I absolutely loved this book. Sure to become a classic. Highly recommend."

— ERIKA BAKKUM, BOOK REVIEWER (US)

"Stoddard takes on the Herculean task of elucidating Einstein's Theory of Relativity, a subject known for its complexity and depth, and accomplishes this with remarkable simplicity and elegance. Whether you are a curious novice or a seasoned science enthusiast, this book will expand your horizons, ignite your imagination, and leave you with a deeper appreciation for the wonders of our universe. The journey through Einstein's universe with John Stoddard is not just enlightening; it is truly transformative."

— SCOTT B. ALLAN, BOOK REVIEWER (US)

"Well-researched, entertaining, vivid story-telling. *Relativity for Beginners* is an enlightening journey through Einstein's groundbreaking theories that revolutionized our understanding of the universe. As an admirer of Albert Einstein's genius, I found this book to be a treasure trove of knowledge. It masterfully simplifies complex concepts, making them accessible to anyone curious about the mysteries of space and time."

— L. WALKER (US)

". John Stoddard makes relativity understandable to the common man and he writes in a style that is fun and engaging. What I like is how he humanizes some of the greatest minds in science. You see them as real people who happened to be extremely gifted and then he makes their discoveries understandable to the average reader. To put it another way, he takes the delicious cookies of scientific discovery and puts them on the lower shelf where the rest of us can reach them. That in itself is genius. This is a book you definitely want to read."

— MARC DE WEBB (US)

"Easy to read and follow along. An easy-to-read book on the seminal work of Einstein and some other greats in the field. Yes, I am guilty of being one of those for whom math and higher-order number crunching were far from my best talents. Very much so. However, this book covers Einstein's journey from patent clerk to college professor and from there to worldwide fame as an original thinker, upending our entire understanding of space, time, and their relationship to each other. This was not a boring book at all. It explained these concepts in a way that I could follow. Well worth reading."

— MARC HENDERSON (US)

"What a fascinating book. Incredible work. This book was written for people like me. An average Joe. The subject matter gives us a better understanding of the science behind relativity. In common language and illustrations. While also blending in a biography of the struggles and successes of Albert Einstein. I was amazed at myself for loving this book so much. Couldn't put it down."

— T.E. MCGEORGE (US)

"*Relativity for Beginners* is a fantastic book that makes understanding the universe's secrets easy and exciting. Whether you're new to these ideas or just want a refresher, this book is perfect. It's like a thrilling adventure that helps you appreciate Einstein's genius and the wonders of space."

— JOHN SPENDER (US)

"Relativity decoded. The book meticulously guides readers through the special and general theories of relativity, unraveling the secrets behind time dilation, the curvature of spacetime, and the famous $E=mc^2$ equation. What sets this book apart is its ability to explain these profound ideas in plain English, making it suitable for both novice enthusiasts and those seeking a deeper understanding.

Relativity for Beginners is a must-read for anyone captivated by Einstein's genius and the profound impact of his theories on our comprehension of the cosmos."

— DAN. W. (US)

"A thrilling unraveling of Einstein's Relativity, suited for those of us mystified by the convolutions of advanced science, bringing life to complex concepts through engaging tales of bizarre discoveries."

— DR. S. W. (US)

Copyright © 2023 John Stoddard. All rights reserved.

The content contained within this book may not be reproduced, duplicated, or transmitted without direct written permission from the author or the publisher.

Under no circumstances will any blame or legal responsibility be held against the publisher, or author, for any damages, reparation, or monetary loss due to the information contained within this book, either directly or indirectly.

Legal Notice:

This book is copyright protected. It is only for personal use. You cannot amend, distribute, sell, use, quote, or paraphrase any part, or the content within this book, without the consent of the author or publisher.

Disclaimer Notice:

Please note the information contained within this document is for educational and entertainment purposes only. All effort has been executed to present accurate, up-to-date, reliable, and complete information. No warranties of any kind are declared or implied. Readers acknowledge that the author is not engaged in the rendering of legal, financial, medical, or professional advice. The content within this book has been derived from various sources. Please consult a licensed professional before attempting any techniques outlined in this book. By reading this document, the reader agrees that under no circumstances is the author responsible for any losses, direct or indirect, that are incurred as a result of the use of the information contained within this document, including, but not limited to, errors, omissions, or inaccuracies.

CONTENTS

Introduction   xiii

Part I
DISCOVERY I

1. Is My Coffee Moving?   3
2. Maxwell Sheds New Light On The Problem   12
3. The Invisible Ether and The Great Clash   16
4. Einstein Rides a Light Beam and Leads a Revolution   20
5. Einstein Follows a Light Beam, Shows Time Is Relative   32
6. One of You Is Lying!   41
7. Honey, I Shrunk the Kids (By Launching Them At Half Lightspeed)   44

Part II
DISCOVERY II

8. On the Shoulders of Giants   51
9. Einstein Stumbles on World's Most Famous Equation   55
10. A Mad Scientist Discovers the Missing Link (And the Deadliest Thing in the Universe)   61
11. The Road to Mankind's Most Dangerous Weapon   70

Part III
DISCOVERY III

12. Einstein's Old Professor Stumbles onto Spacetime   79
13. Problems with Newton's Gravity   85

14. Einstein Day Dreams About Falling, Has the Happiest Thought Of His Life — 90
15. Show Me The Light — 95
16. Gravity Warps Space — 104
17. Einstein's Old Friend Shows Him the Way — 108
18. Einstein's Great Masterpiece Revealed! — 120
19. We Need Proof! — 126
20. So What? — 130
21. Unknown Genius Solves Einstein's Equations, Discovers A Miracle — 134
22. The Teacher, The Janitor, The Priest, The Russian, and the Patent Clerk — 144
23. In The Beginning — 151
24. The Dark Side Of The Force — 155
25. The Search for A Theory of Everything And Final Thoughts — 160

Acknowledgements and Reviews — 165
Author's Other Works — 166
Glossary — 167
Quick Reference Guide — 179
Equations — 182
References — 189

For Ashley

INTRODUCTION

"Nothing goes over my head. My reflexes are too fast. I would catch it."

— DRAX THE DESTROYER, GUARDIANS OF THE GALAXY (2014)

"Newton was wrong about Time! There's no time to explain, I'll tell you everything tomorrow."

Maurice Solovine watched his friend dash out of the café with skeptical eyes. "Who does this crazy violin playing patent clerk think he is?" he thought to himself. "Why won't he take his head out of the clouds and leave physics to the physicists?"

Still, Solovine had a special fondness for his friend, but even he was forced to admit his friend didn't inspire much confidence. It had taken his friend two tries to get into the Polytechnic Institute, and once he finally made it in by the skin of his teeth, he spent more time smoking and skipping class than he did attending lectures, so much so that his professors refused to offer him a returning role as a professor following graduation. Outcast from the university, his friend had been forced to take on a menial job as a patent clerk to support his wife and newborn son. Still, there was something about this maverick, something about his wild hair or rebellious spirit, that Solovine couldn't help but admire. "Talk soon, friend," Solovine muttered to himself. Solovine drained his coffee, left a tip, and slipped out of the café.

The year was 1905. Though no one could have ever predicted it, Solovine's friend, this simple patent clerk who couldn't even speak full sentences until he was four years old, would publish four papers later this year, his "Miracle Year," and would go on to fundamentally alter our understanding of gravity, light, matter, and reality itself.

Though he would never run for public office or lead an army, his secret letter to President Roosevelt would change the outcome of WWII and inspire the invention of the most dangerous weapon mankind has ever devised.

Though he would never witness modern GPS satellites or lasers, his ideas would inspire their creation as well as countless other inventions.

Though he never saw a black hole, his theories predicted their existence decades before we photographed one.

Even in his own lifetime, the world would regard him as one of the greatest minds our species has ever produced, in the ranks of Galileo, Newton, Voltaire, Plato, and Aristotle. In time, his fame would grow so great the people would shout his name in the streets wherever he went.

His name was Albert Einstein.

In a strange twist of irony, this peculiar man, one of the greatest Thinkers in all human history, had a deep passion to understand the secrets of the universe, but he struggled in school (perhaps some of us can relate). Indeed, Einstein himself despised the German school he attended as a youth because it prized rote memorization of facts and figures above true understanding or mastery, and because it stole precious time away from him to do that which he loved more than anything else in the world, which was to *think*. Incidentally, not overly much thinking is encouraged in modern schools either. Freedom of thought, imagination, and debate (the tools of the Thinker) have been traded in for productivity, efficiency, and obedience (the tools of the Worker): thus, there is little wonder why our youth have lost their wonder.

Unlike Einstein, I'm not a world renowned theoretical physicist and am just a humble engineer by trade (and a third-rate violinist at best), but like him I'm prone to look at the stars with curious eyes. Like him, I stand in awe of the Universe and desire to learn her secrets, and so over a period of several years after reading hundreds of books and journals through college and beyond, I've been able to move forward in my journey of understanding, inch by inch, and this journey has brought me indescribable joy. And I believe that all citizens of the world deserve to experience this joy, whether they have the time or resources to obtain an advanced degree or not.

By the time you're done reading this short volume, you will absolutely know more about Special Relativity and General Relativity than 99% of the population (and certainly more than the writers from *Marvel* or *Back to the Future*). Even better, you will understand the biggest problem facing physicists today – the struggle to uncover a Theory of Everything, to unite Classical Physics and Relativity with Quantum Physics.

Along the way, you will discover:

- **The hidden secrets of Space and Time.**
- The solar eclipse that saved Einstein's career.
- **The true secret nature of light (it's too strange to believe).**
- How a playful riddle about a dead cat fundamentally changed our perception of reality itself.
- **The #1 deadliest killer in our universe (and why we're still alive).**
- Einstein's top-secret letter to President Roosevelt that helped the Allies win WWII.
- **The ONE thing in our universe that moves faster than light.**
- The secrets of worm holes, black holes, multiple realities, and String Theory (and what's wrong with it).
- **How a genius you've never heard of took us to the brink of a Theory of Everything. And a whole lot more!**

So, just as Einstein did throughout his life, I invite you to awaken your Inner Child to the wonders and majesty of the strange and bizarre universe we live in, as we explore the secrets of Relativity together.

Because the answers to the greatest mysteries of the universe, as Einstein often discovered, could not be discovered with fancy calculators or telescopes or microscopes or rulers.

Instead, the truth, Einstein often discovered, could only be unlocked from within.

From the power of one's own free-wheeling, childlike imagination.

PART I

DISCOVERY I

EINSTEIN DAY DREAMS ABOUT RIDING A LIGHT BEAM & LEADS A REVOLUTION

1

IS MY COFFEE MOVING?

"The only thing that ever got in the way of my learning was my education."

— *ALBERT EINSTEIN*

Somewhere inside a grey school building located on a spinning, watery rock orbiting a tiny yellow sun on the edge of a spiral arm of a spinning galaxy orbiting black hole Sagittarius A, my old physics professor took a sip of coffee from his favorite mug, set it down on the desk, surveyed us, an advanced class of sixty students, and asked a simple question. I didn't know it at the time, but this innocent question would haunt me for years to come.

"Is my coffee moving?" he asked.

We eyed each other nervously. Was this some kind of a joke? No one spoke as the seconds ticked by.

Could it be? Finally, here was a question I could answer with some confidence. I raised my hand to answer and felt the eyes of the class, including the beautiful green eyes of my crush sitting two rows in front of me, settle on me. This was my moment.

"No, Professor, your coffee isn't moving."

The professor shook his head and smiled.

"John...Not only are you wrong, but I've never been more disappointed in your lazy thinking than I am right now. When will you learn to see with more than just your eyes? Is there anyone else in this class who isn't absolutely ignorant?"

Seconds became minutes as we sat in silence and I unsuccessfully attempted to melt into the floor.

Finally, a shy student in the back quietly said, "Yes, Professor, the coffee is moving."

"Explain yourself."

"Well," she hesitated. "The coffee isn't moving to us."

"Go on," the professor said as the beginning of another smile flickered on his face.

The coffee isn't moving to us, but the coffee is located on our planet, which is rota-"

"Rotating at 1,000 miles per hour every second," the professor cut her off. He was fully beaming now.

"The coffee is attached to our planet which is rotating and which will continue to rotate long, long after our species survives the next mass extinction event, dies in it, or flees to a different planet entirely. Additionally, the planet is orbiting the Sun at 67,000 miles per hour, and on top of that, our entire solar system is inside of our Milky Way galaxy, which is spinning at 130 miles per second. And as if that isn't enough, the fabric of the universe itself is expanding faster than the speed of light, so fast that even if intelligent civilizations existed elsewhere in the cosmos (and they probably do), all of them except the very closest ones would have to break the lightspeed barrier just to make it to our planet, which is itself a moving target."

Fig. 1: Where Am I? "Milky Way". Image Credit: Pablo Carlos Budassi, licensed under CC By 4.0.

The professor paused briefly to gaze at us again as we listened to him, hanging on every word.

"These are the questions the gold conjuring virgin Isaac Newton and the great college dropout Galileo himself grappled with. The truth is that everything in our cosmos is in constant motion, and there is no such thing as 'absolute rest.' And so therefore you can never say my coffee isn't moving. All you can say is that my coffee isn't moving relative to you."

The professor paused one more time to let the full weight of this truth sink in.

"Class dismissed."

400 years ago in Florence, Italy, before we ever knew the Earth was spinning or that our Milky Way galaxy was just one of countless other galaxies hurtling through space, a man named Galileo Galilei was standing at the harbor watching a ship glide through the water when he was struck by a strange and curious thought: *"If I was standing on the ship beneath the deck with no windows moving at constant speed, would I be able to tell I was on a moving ship?"*

Galileo continued his thought experiment. *"And if I dropped something while standing inside the moving ship, wouldn't it appear to drop straight down to me?"* In a final lightning burst of inspiration, Galileo thought, *"And if I did peer out the window on board the ship, would I be able to tell that I am moving through the water toward the harbor, or would it not appear as though I were the one standing still and the harbor was moving toward me instead?"*

In his masterpiece *Dialogue* in 1632, Galileo gave us his **dictum on uniform motion**, which holds that the laws of physics for **reference frames**, or points of view, moving at constant or uniform speed relative to each other are the exact same as the laws of physics for reference frames at rest relative to each other (Egdall, 16). In other words, if you're trapped below the deck of a ship moving at 5 miles per hour with no windows (or a space ship travelling at 5 million miles per hour through space with no windows), *you would never be able to tell you were moving at all.*

Strange Facts: *Galileo would then make the intellectual leap with Copernicus that would cost him his freedom and proclaim that if the Earth was in fact moving, we would never be able to tell because we ourselves inhabit its surface. The Catholic Church, which insisted the Earth was the non-moving center of the Universe, placed the seventy-two year old Galileo under house arrest in 1633 for heresy, and his beautiful masterpiece, Dialogue, was banished to the Index of Prohibited Books for the next 200 years. Galileo died in 1642, and today his middle finger is (deliciously) displayed inside the Museo di Storia del Scienza in Florence, Italy for all to see.*

Fifty years or so later in 1687, the great Isaac Newton expanded on Galileo's ideas and published what is regarded as one of the most important works ever produced by mankind, his *Principia*. In addition to defining space and time and gravity, Newton gave us his three famous, world-changing principles governing how forces affect the motion of objects. We know these three principles today as **Newton's Laws of Motion**, which are defined briefly here:

1. An object in motion will stay in motion until acted upon by an outside force. An object at rest will stay at rest until acted upon by an outside force.

2. Force is equal to the mass of an object times its acceleration, the rate at which it changes speed or direction.
3. For every action, there is an equal and opposite reaction.

At this point, careful readers may be asking a very good question, the same question Newton's peers asked him even in his own time, even before they were able to prove he was correct through observation - do Newton's Laws obey Galileo's Dictum?

Through intuition and experience, we know Newton's Laws hold true in all reference frames moving in uniform motion relative to each other (a bouncing ball behaves the same when you're standing still as it does when you're standing on a moving train), but it would take another *200 years* before physicists were finally able to prove *mathematically* that Newton's Laws obey Galileo's Dictum, through a set of equations that are known today as the **Galilean Transform** (Egdall, 22). Though it is beyond the scope of this book to derive these equations in detail, the Galilean Transform shows us exactly how to transform the coordinates of an event in any given reference frame in uniform motion to the correct coordinates of any other reference frame in uniform motion, and when we plug in Newton's Laws into the Galilean Transform, they remain unchanged – Newton's

Laws do, in fact, obey Galileo's Dictum. They are *not* affected by uniform motion.

MYSTERIOUS FORCES AND THE PESKY PROBLEM WITH MAGNETS AND ELECTRICITY

Newton's Laws dominated scientific thought on force and motion until 1820, when a Danish scientist named Hans Christian Ørsted placed a wire next to a magnetic compass, ran a current through the wire, and *saw the compass needle move*. Newton's Laws described how forces affect the motion of objects when they collide with other objects – *but what about when the objects aren't touching?* Incredibly and mysteriously, the current running through the wire was applying force through empty space to move the needle of the compass. And just eleven years later in 1831, the great thinker Michael Faraday demonstrated the complete opposite effect – when he moved a magnet near a bit of wire, he was able to induce an electric current (Belendez, 2015).

So Ørsted was able to show how an electric current can produce a magnetic field (the basis of the modern motor), and Faraday was able to show how a magnetic field can produce an electric current (the basis of the modern generator), and by the mid-1800s, they along with several other pioneers had even crafted several mathematical equations to describe magnetic and electric phenomena. But there were several problems with the equations – first, the equations violated the law of conservation of charge, which holds that electric charge in a closed system always remains the same. Even worse, the equations represented a random collection of separate theories, and no one could mathematically describe yet exactly how electricity and magnetism were connected, even though the connection was obvious from experimental observation. Unfortunately, Newton's Laws weren't equipped to describe the strange new science of electricity and magnetism either.

Another genius was needed to bridge the gap – that genius, as fate would have it, turned out to be Faraday's own student, the greatest physicist you've never heard of.

2

MAXWELL SHEDS NEW LIGHT ON THE PROBLEM

"What if I could ride a beam of light across the Universe?"

— ALBERT EINSTEIN

"I'm sure the universe is full of intelligent life. It's just been too intelligent to come here."

— ARTHUR C. CLARKE

The Greatest Physicist You've Never Heard Of, the Father of Electromagnetism whom Einstein himself regarded as even greater than Sir Isaac Newton, the unsung hero who laid the groundwork for every bit of Einstein's contributions to Science, was none other than Michael Faraday's humble student and lab assistant, the Scottish inventor-philosopher-poet James Clerk Maxwell. Maxwell, propelled forward by Faraday's work with electricity and magnetism, carried the torch further and united the separate theories of electricity and magnetism together with just four simple equations, four partial differential equations which concisely and accurately described the relationship between electricity and magnetism – namely, that a changing electric field produces a magnetic field and that a changing magnetic field produces an electric field.

But his most brilliant insight was that these changing electric and magnetic fields not only produce each other but are *coupled* together to produce **electromagnetic waves** that propagate through space. The simple motion of an electron is enough to produce such a wave: when the electron moves up and down, it produces a changing electric field because it has a charge, which in turn produces a changing magnetic field. These fields run concurrently and perpendicularly to one another in a never-ending, self-perpetuating cycle through space.

Maxwell reasoned the motion of the electron through space drives the electromagnetic wave that is produced, and if the electron ever stopped moving, the electromagnetic wave it creates would immediately cease to exist as well, since the wave relies on the relative motion of the electron. The wave *must* move to exist. *Maxwell didn't know it at the time, but this singular idea, simple on its surface, would inspire Einstein to discover Relativity just forty years later.*

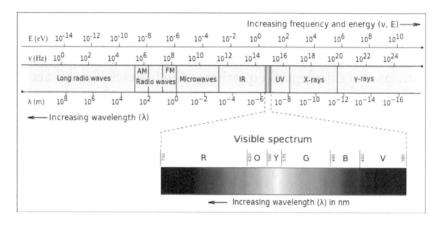

Fig. 2: Electromagnetic Spectrum. Visible light, gamma rays, infrared waves, radio waves, and ultraviolent waves have different wavelengths and frequencies, but they are all a form of light. Image Credit: Philip Ronan, licensed under CC By 4.0.

But Maxwell didn't stop there. Through experimental measurement of the observed wavelength and frequency of electromagnetic radiation, he deduced the velocity of electromagnetic radiation to be exactly equal to the speed of light. Therefore, Maxwell reasoned, all visible light is nothing more than a particular *kind* of electromagnetic radiation.

The different kinds of electromagnetic radiation include gamma rays, x rays, infrared rays, microwaves, radio waves, and visible light, but *all* of them are a form of light.

Notice that according to the figure above, as a wave's frequency increases, so does its energy. This idea will become important again when we discuss General Relativity.

By the dawn of the 20th century, Ørsted and Faraday had fathered the Industrial Age with the motor and the generator, and Maxwell had carried the torch one step further by uniting the laws of electricity and magnetism into a single, spectacular overarching theory. The Communication Age was born (O'Callaghan, 2021).

The exciting technologies stemming from Maxwell's new science of electromagnetism, unfortunately, were outpacing the theoretical understanding *behind* the science – in a David and Goliath-style battle, Maxwell's ideas were about to clash directly with those of Newton and Galileo, spark a revolution, and challenge over 200 years of accepted thought about the nature of Space and Time itself.

3

THE INVISIBLE ETHER AND THE GREAT CLASH

"No great mind has ever existed without a touch of madness."

— ARISTOTLE

By the turn of the century, physicists wrestled with two major problems with Maxwell's equations.

One, Maxwell's equations did not obey Galileo's Dictum, which, we recall from Chapter 1, tells us the laws of physics for all reference frames in uniform motion are the same. When physicists plugged Maxwell's equations into the Galilean Transform, as they had done with Newton's Laws,

Maxwell's equations didn't stay the same – they changed. This test indicated that electricity and magnetism *are* in fact affected by uniform motion.

That said, this conclusion did not seem to line up with experience in the real world. For example, physicists were well aware that an electric motor or generator or radio (or any technology that relies on electromagnetism to function, for that matter) is equally capable of functioning on a train moving at constant speed as it is capable of functioning at rest. Perhaps something was wrong with Maxwell's equations? Or perhaps something was wrong with the Galilean Transform? *Either one or the other was wrong.*

Intense debate among the leading physicists in the late 1800s sparked even deeper questions and exposed the second major problem – namely, Maxwell's equations insisted that light always travels at constant speed. But what does this mean in the real world? Light moves at the same speed relative to *what* exactly? *These* questions in particular forced physicists to take a hard second look at one of their most sacred and time-honored theories, the theory of the **Ether**.

THE INVISIBLE AND MYSTERIOUS ETHER

Since the time of Newton, scientists recognized that water waves propagate through water and sound waves propagate through the air – what did light waves propagate through? To answer this question, Newton surmised that light waves travel through the "luminiferous Ether," a mysterious background substance that permeates space and is *the* standard of absolute rest. Earth, according to Newton, moved along with all the other planets through the luminiferous Ether, as did sunlight. When light waves moved in the direction of the Earth's movement through the Ether, Newton predicted the light's speed would decrease as it faced an Ether "headwind," just as you feel your hand blown back by the wind when you stick your hand out the window while travelling in a car.

In 1887, physicists Albert Michelson and Edward Morley devised an ingenious way to test the Ether theory and prove its existence once and for all by setting up a light beam "race" with a clock and mirrors. One beam of light would travel in the direction of the Earth's movement (against the proposed Ether) and back while the second beam would travel the same distance in a perpendicular direction and back. The Ether, they hypothesized, would cause the first beam to slow down on its journey leg against the Ether and lose the "race," not too differently from the way a swimmer fighting directly against the current will travel more slowly than a swimmer cutting *through* the current.

But Michelson and Morley's light race experiment produced shocking results – according to their instruments, the light beams tied every single race. They tried several variations – in one variation they even waited six months for the Earth to move in the opposite direction relative to the sun - but no matter what they did, the light beams always travelled at the exact same speed. In other words, there was no observable evidence the Ether existed at all (Lincoln, 2022).

To make matters worse, Michelson and Morley's experiments, which perfectly supported Maxwell's assertion that light always travels at the same speed, did nothing to stop the physics community from searching for the Ether. Unfortunately, both Newton and Maxwell's followers *needed* the undiscovered Ether to be real because it was the ultimate standard of absolute rest. *To say that light always travels at the same speed only makes sense if we say it is moving at the same speed relative to something else that is always at absolute "rest" – therefore the Ether must be real.* A stubborn, illogical obsession with the Ether had gripped the physics community as they rushed to find other "better" ways to detect it, never stopping to question if it truly existed in the first place.

The answer to this riddle would ultimately come from the unlikeliest of places, far, far away from the prejudices of the self-important universities and lecture halls, from a lowly twenty six-year-old patent clerk with a talent for mischief and day dreams.

4

EINSTEIN RIDES A LIGHT BEAM AND LEADS A REVOLUTION

"To punish me for my contempt for authority, fate made me an authority myself."

— ALBERT EINSTEIN

Sixteen-year-old Albert Einstein roamed the Italian countryside on his bicycle, blissfully losing himself in the endless corridors of his own wild imagination. The year was 1895 and it had been a tumultuous one for young Albert. A year earlier, he had been expelled from the suffocating, militaristic Gymnasium school in Munich, Germany for unruly and disrespectful behavior toward his teachers. His father had left him behind in Munich to try opening another electrochemical business in Pavia, Italy.

Following his expulsion, Albert simply hopped on a train across the Alps to rejoin his family in Italy, where his parents welcomed him with a funny mixture of disappointment and awe of their boisterous child. Albert simply couldn't be tamed.

Here in the beautiful Italian countryside, Albert felt alive and free, and his new school encouraged him to make full use of his imagination, unlike the strict German school which only valued rote memorization of facts and figures. Einstein was pedaling happily through his favorite trail when a tantalizing question popped into his head that would haunt him for years to come.

"What would happen if instead of sitting on a bicycle, I was sitting on a beam of light?"

Even then, Albert was already familiar with the work of Galileo and Newton and the ideas of relative motion. "If indeed I could move as fast as a beam of light," Albert reasoned, "wouldn't the light wave appear to stand completely still relative to me?"

Ten years later on a lazy afternoon in Germany, Albert Einstein, now a young man, sat at his desk where he worked as a patent clerk and, pushing his patent work to the side (he had already finished the day's work two hours before lunch time), he pulled out his secret scribblings and notes to ponder the very same question that had plagued him as a teenager in Italy, the question of the light beam.

"We know that light is an electromagnetic wave that must move in order to exist," he thought.

"Therefore, it must be true that no matter how fast I move, I will never be able to catch up to the light because if I travelled at the same speed as the light beam, it would cease to move relative to me and would therefore no longer exist to me, which can't be true.

"And if I can't catch up to light no matter how fast I move, light must be the fastest thing in our universe.

"We also know Maxwell's Equations show us that light's speed never changes," he thought.

Here, a rebellious and playful thought formed in his mind:

"What if the Ether does not exist? What if there is no such thing as absolute rest and all things are in constant motion relative to each other? According to the evidence, the theory of Ether (and absolute rest) is 'patently' false."

Einstein chuckled to himself and drew from his pipe. And here, the thought that would change the world struck him like a lightning rod:

"What if Maxwell is to be taken seriously, and the speed of light is absolutely the same in each and every situation, inside all reference frames and for all observers, no matter how fast or slow the observers are moving relative to anything and everything else?

"If we accept this to be true, what would be the consequences?"

Later that year in 1905, Einstein published his landmark paper on **Special Relativity**, entitled "On the Electrodynamics Of Moving Bodies," and the world would never be the same.

Strange Facts: After Einstein's death in 1955, Thomas Harvey, the man in charge of his autopsy, secretly stole Einstein's brain. After dissecting it into over 170 pieces, Harvey sent several pieces to select neuroscientists and pathologists for study – what they found was remarkable. The structures in the back and top region of the brain which are responsible for spatial and mathematical reasoning were 15% larger than average. Einstein's family eventually learned about the theft but allowed Harvey to continue his work, provided that he promised not to leak the scandal to the media. A reporter tracked Harvey down nearly twenty five years later to inquire about Einstein's missing brain and pulled the story out of him.

ARGUING WITH NEWTON'S GHOST

To understand exactly how bizarre and revolutionary Einstein's thought was – that the speed of light is absolutely the same for all observers in all reference frames – it's helpful to first revisit the physics of Newton, the physics of what we see in the real world for objects travelling much, much slower than the speed of light.

As a simple test, let's first imagine a baseball pitcher Dave standing on the mound, winding up for another fastball. He throws his fastball and the clock reads "40 feet per second."

Next, let's place Dave on a train which is moving at 50 feet per second. Dave winds up for another fastball and releases it at the exact same speed as his previous pitch. What's the speed of the fastball?

According to Newton, the clock on the train would read "40 feet per second." Newton's Laws obey Galileo's Dictum, and the ball behaves the same way inside the moving train as it does on the baseball diamond. But imagine Dave's friend Nolan is standing outside of the train watching it roll past at 50 feet per second – how fast is Dave's fastball to Nolan?

According to Newton, if you're standing outside the train as it moves by you, you must remember to combine the speed of the fastball with the speed of the train to calculate its relative velocity to you. As an outside observer Nolan would measure a speed of 90 feet per second – the velocity of the fastball inside the train (40 feet per second) plus the velocity of the moving train (50 feet per second) for a total of 90 feet per second.

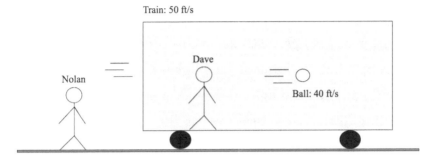

Fig. 3: Relative Velocities on a Moving Train. Image Credit: Author's work.

Here's Newton's speed combining formula, which is very simple.

For speeds v_1 and v_2, the combined speed is:

$$V = v_1 + v_2$$
$$V = 50\frac{ft}{s} + 40 ft/s$$
$$V = 90\ ft/s$$

Eq. 1: Newton's Speed Combining Formula

And this makes intuitive sense. The observer outside the moving train will see a much faster pitch than the observer inside the moving train.

Now *here* is where things get interesting and Einstein's troublemaking spirit comes into play:

Imagine Dave is standing inside a train which is moving at half the speed of light, .5c, except this time instead of holding a baseball he's holding a flashlight. On the signal, Dave switches on the flash light and sends a horizontal light beam travelling in the same direction as the train. How fast does the light travel, according to Dave on the train and according to Nolan, the observer standing outside of the moving train?

First let's take a look at what Newton would say. Newton would say that for Dave on the train, the light beam would simply travel at the speed of light c. From Nolan's point of view outside of the train, Newton would insist that the speed of light c must also be combined with the speed of the moving train, .5c:

Velocity of the light from the standing observer's reference frame:

$$V = v_1 + v_2 = 0.5c + c = 1.5c$$

But Einstein insists this is impossible, since 1) nothing can ever move faster than light, and 2) Maxwell's equations say that light speed is always the same, and therefore the observer outside the train must *also* see the light beam travel at velocity c. According to Einstein, Newton's speed combining formula is a rough approximation that only works for relatively slow-moving objects. For objects travelling close to the speed of light, in order to establish light as the fastest moving thing in the universe and to make it constant in all reference frames, Einstein was forced to invent a *new* speed combining formula that perfectly matched Newton's results for slow moving objects but *also* accounted for objects moving at speeds approaching that of light:

For speeds v_1 and v_2, measured as a percentage of the speed of light, the combined speed V is:

$$V = \frac{v_1 + v_2}{1 + v_1 v_2}$$

Eq. 2: Einstein's Speed Combining Formula (Einstein, 29)

Let's take one more look at Dave's flashlight experiment from the standing observer's reference frame with Einstein's formula:

For two initial speeds, $v_1 = 0.5$ (as a percentage of the speed of light c) and $v_2 = 1.0$ (as a percentage of the speed of light c), the combined speed V is:

$$V = \frac{v_1 + v_2}{1 + v_1 v_2}$$
$$V = \frac{0.5 + 1.0}{1 + (0.5)(1.0)}$$
$$V = \frac{1.5}{1.5}$$
$$V = 1$$

In other words, the speed of the light from the standing observer's reference frame is 1 or 100% of the speed of light, c.

According to Einstein, when objects move close to lightspeed, Newton's theories no longer work. Both the standing observer Nolan and Dave on the train see light travelling at the same speed c, just as Maxwell's Equations declare (Egdall, 55).

Here, careful readers may recall that Maxwell's Equations were hotly debated among academic circles because when physicists plugged them into the Galilean Transform, they changed, thus violating Galileo's Dictum that the laws of physics work the same in all uniformly moving reference frames. But when Einstein published his 1905 paper on Special Relativity, he derived a *new* transform to be used for Maxwell's Equations, and when he plugged Maxwell's Equations into his new transform, they emerged unchanged.

Thus, he was able to show that Maxwell's Equations apply in all uniformly moving reference frames, just like Newton's Laws of Motion. In other words, light always moves at the same speed relative to you, whether you're standing still or on a train or boat or car moving at constant speed.

Without access to other contemporary physicists and their work, Einstein worked in solitude six days a week in the patent office and independently derived the new transform that showed that Maxwell's Equations are not affected by uniform motion all by himself. Incredibly, the exact same mathematical work had been performed by physicists Hendrik Lorentz and George Fitzgerald years before, but with the wrong purpose – they derived these equations in order to defend Michelson and Morley's failed light race experiment and defend Ether's existence. For this reason, the Lorentz-Fitzgerald transform is more popularly known as the Lorentz-Einstein transform – it was Einstein alone who had the insight to reject the Ether altogether and lead a revolution.

Fig. 4: Young Albert Einstein. Image Credit: Lucien Chavan, licensed under CC By 4.0.

From here, Einstein was able to take Galileo's Dictum a step further and propose his **Relativity Postulate**:

All laws of physics are the same for all uniformly moving reference frames.

At first glance, this is exactly the same as Galileo's Dictum, except that in the time of Galileo, electromagnetism had not been discovered yet – Einstein was saying that *all* laws of physics, including electromagnetism and as yet undiscovered physical laws, are not affected by uniform motion.

And after vindicating Maxwell's Equations with the Lorentz transform, Einstein formally gave us his **Light Postulate:**

The speed of light is absolutely the same in all uniformly moving reference frames, independent of the speed of the light source and independent of the speed of the observer.

The meaning of this world-changing statement, if taken to its logical limits, is stranger than fiction and goes against everything we observe in our normal lives. For example, if you're moving in a car at 90 miles per hour and a car zips by you at 100 miles per hour, you and Newton would say that car is moving at 10 miles per hour relative to you. But light doesn't work like that, Einstein says.

1. Einstein is saying if a car shined a light in your direction, you would measure the lightspeed as c.
2. Einstein is saying if the car raced *toward* you at 500 million miles per hour and shined a light toward you, you would *still* measure the lightspeed as c.

3. Einstein is saying if you, the observer, sprinted *to* the shining light on the car at 500 million miles per hour, you would measure the lightspeed as c, and
4. Einstein is saying if you sprinted *away* from the shining light on the car at 500 million miles per hour, you would *still* measure the lightspeed as c.

The speed of light is always the same for all observers in all uniformly moving reference frames.

If this sounds too strange to believe, you're in good company. But Einstein was just getting started. Realizing that speed is a function of distance through space and time, Einstein realized that in order to keep the speed of light c constant in all reference frames for all observers, then *space and time itself would be forced to bend, stretch, and contract for different observers.*

But didn't Newton say that time and space are absolute? A second for you is the same as a second for me. A meter to you is the same as a meter to me, right?

No, according to Einstein. He was about to show that Space and Time are relative.

5

EINSTEIN FOLLOWS A LIGHT BEAM, SHOWS TIME IS RELATIVE

"Put your hand on a hot stove for a minute, and it seems like an hour. Sit with a pretty girl for an hour, and it seems like a minute. That's relativity."

— ALBERT EINSTEIN

So Time and Space aren't absolute after all, as Newton thought, but instead they're relative. They can bend and stretch and change.

Einstein discovered this, not with fancy mathematics and calculations, but with yet another ingenious day dream – his famous **Light Clock thought experiment**. Imagine there are two stationary parallel mirrors placed vertically on top of each other separated by a foot or so with a photon bouncing between them at the speed of light. The clock ticks and tocks as the photon makes contact with each mirror. This is the stationary Light Clock. Stationary Stacy is standing next to this Light Clock. Now, imagine *another* Light Clock that is the exact same and also has a photon bouncing up and down between the mirrors, but this Light Clock is moving horizontally at a constant speed inside a train. Moving Michael is on board the train watching it closely.

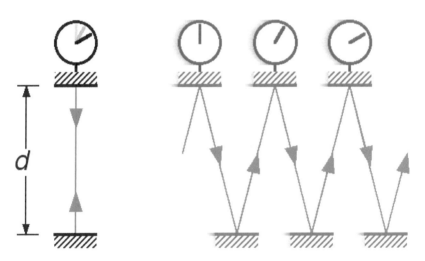

Fig. 5: Light Clock thought experiment. Both observers observe their photon is bouncing up and down for their individual Light Clocks. But an observer at rest relative to a moving Light Clock will observe that the photon of the moving Light Clock must travel a longer distance as it bounces between mirrors which are uniformly moving. Image Credit: Michael Schmid, licensed under CC By 4.0.

What do Stationary Stacy and Moving Michael see and experience? To Stationary Stacy, time is passing normally, and to

Moving Michael, time is passing normally too. When they look at their Light Clocks, all is normal and the photon is bouncing straight up and down between the mirrors.

It is only when either of them looks at the *other* person's Light Clock that things start to get interesting.

When Stationary Stacy watches Moving Michael's Light Clock, she sees the photon moving diagonally up and down as Moving Michael's Light clock moves from left to right and she also notices Moving Michael's Light Clock is ticking slower.

Tick....Tock....Tick....Tock. Einstein tells us that the speed of light is always constant. And since Stacy sees the photon on the moving Light Clock has to travel a greater distance as it moves diagonally up and down (and since we also know that the speed of the photon is the same inside both Light Clocks), then we also know that Stacy will observe the photon on Moving Michael's Light Clock takes longer to reach the mirror than the photon on her own Light Clock. In other words, she will observe that photon bounces occur faster on her own Light Clock than photon bounces on Moving Michael's Light Clock. *Once again, she will observe that photon bounces occur faster on her own Light Clock than photon bounces on Moving Michael's Light Clock.* In other words, time is ticking faster for Stacy than it is for Michael.

So to Stacy, Moving Michael's Light Clock is ticking more slowly than her own Light Clock – time is slower *inside the train*. Moving Michael's time is moving slower than Stacy's **proper time**, her "wristwatch time" measured inside her frame. While Stacy hears her Light Clock ticking normally (ie, tick-tock-tick-tock), she hears Michael's Light Clock ticking more slowly: tick....tock....tick....tock.

The exact same effect is observed from Moving Michael's frame of reference. When he looks outside the train at Stationary Stacy's Light Clock, he perceives that *he* is stationary and *her* Light Clock is rushing past him as the train moves. He watches the photon inside her Light Clock follow a diagonal path as her Light Clock appears to rush past him. He also sees that *her* Light Clock is ticking more slowly than his Light Clock – once again, since the speed of light c must be constant in both Light Clocks, a greater observed distance travelled by Stacy's photon requires a greater time duration per photon bounce, and so he perceives Stationary Stacy's time is moving slower than *his* proper time, his own "wristwatch time."

Both Stationary Stacy and Moving Michael are correct. Einstein tells us both frames of reference are equally valid.

When an object moves, time slows down for the object relative to stationary observers (Einstein, 26).

With just a bit of algebra and the Pythagorean theorem, Einstein was able to derive an equation to calculate the exact amount of **time dilation** Moving Michael and Stationary Stacy would experience using this exact same light clock thought experiment and following the path of the diagonally moving photon:

$$\Delta t' = \frac{\Delta t}{\sqrt{1 - v^2}}$$

Eq. 3: Einstein's Time Dilation Equation, where:

$\Delta t'$ = dilated time of an event observed by someone in a different reference frame

Δt = time observed for an event in one's own reference frame (rest time)

v = speed of the moving reference frame as a percentage of the speed of light

$\sqrt{1 - v^2}$ = Lorentz Factor

Notice, once again, the genius of Einstein's equation. At normal life speeds, the speed of a moving reference frame is much, much lower than the speed of light (practically zero), and so any time interval measured by a moving observer's wristwatch and a resting observer's wristwatch would be very nearly *identical*. In Newton's world, a second for one observer is the exact same as a second for a different observer. It's only when we deal with speeds approaching the speed of *light* that time dilation becomes a very real and serious problem (Szyk, 2023). Let's walk through a quick thought experiment to show this miraculous science in action:

Iron Man and Thor are in another heated argument that threatens to break up the Avengers forever. To cool off, Thor hops in a space ship travelling 90% the speed of light in the direction of Asgard. After one hour of space ship time has passed according to Thor's watch, how much time has passed for Iron Man on Earth (assuming Earth is at rest relative to Thor's space ship)?

If Newton were alive, he would say one hour.

But Newton can't help us here.

So in order to calculate the dilated time for Iron Man we must plug in Thor's proper time into Einstein's time dilation equation:

$$\Delta t' = \frac{\Delta t}{\sqrt{1-v^2}}$$

$$\Delta t' = \frac{1hr}{\sqrt{1-(0.9)^2}}$$

$$\Delta t' = \frac{1hr}{0.4346}$$

$$\Delta t' = 2.301 hr = 2hr, 18min, 5sec$$

Let's pause to consider how incredible this is. One hour in Thor's space ship time is equal to over *two hours* in Iron Man's Earth time. In other words, Thor's space ship time is moving more slowly than Iron Man's Earth time. Assuming Thor keeps travelling this fast, ten years in space ship time would equal just over twenty-three years in Earth time. If you've ever seen the movie *Interstellar*, you will recall exactly how incredible (and dangerous) time dilation can really be.

When Einstein published his revolutionary paper and described how motion causes time to slow down relative to stationary observers, his ideas were so far ahead of his time, his contemporaries were unwilling or unable to properly appreciate it - he faced intense criticism and ridicule from his peers. One of the most famous attempts to prove Einstein wrong was the so-called **Twin Paradox**, and a version of it is given here:

Suppose there are two twins, Earthly Earl and Rocket Ralph. Earthly Earl stays on Earth while Rocket Ralph leaves in a space ship to a distant planet at half the speed of light, then turns around and returns to Earth. Which twin is younger when Rocket Ralph lands on Earth?

Rocket Ralph is moving, so therefore time slows down for him compared to Earl, and so when he returns, he will be younger than Earthly Earl. *But* relativity works in both directions. From Rocket Ralph's perspective, he is stationary and when he leaves Earth, he sees *Earthly Earl* moving away from him at half the speed of light, and when Rocket Ralph returns, he perceives that Earthly Earl is moving *toward* him at half the speed of light. To Rocket Ralph, *he* is the stationary twin, and Earthly Earl appears to be moving, and so therefore time moves slower for Earl than it does for Ralph, and *Earl* will be younger.

Both twins insist the other should be younger, but that's impossible. Finally, Einstein is stumped.

As usual, Einstein had an answer for this thorny riddle – time dilation applies to both observers as long as they are in *uniform* motion relative to each other, meaning there is no **acceleration**. In this example, Rocket Ralph experiences rapid deceleration when he stops at the distant planet, and more acceleration (change of direction) when he turns around to return to Earth. The act of *acceleration* causes the relativistic effects to apply only to him – when he returns to Earth, Rocket Ralph will be the younger twin (Lasky, 2003).

Incredibly, time dilation occurs all around us every day for all objects in motion, though it's extremely difficult to notice at normal, everyday speeds that are much, much slower than the speed of light. When you watch cars zip by you, you would measure time moving more slowly inside the moving car than it does on your wristwatch. But the difference is so small it's practically impossible to tell the difference.

If Einstein had stopped *here* in his writings, he would have already accomplished more in a single paper than most scientists accomplish in their entire lifetime. In only a few pages, he had overturned two centuries of Newtonian thought and shown that Time is not absolute. And this paper was only one of three he published in June 1905 – though it is beyond the scope of this text to explore these topics in depth, in the other two papers he showed that light has a particle nature (co-founding Quantum Physics), and he calculated the size of atoms. But as is typical of Einstein, he was nowhere close to being finished, not by a long shot. In the next chapter, we'll continue down the rabbit hole of Time and explore one of the most bizarre paradoxes in all of physics.

6

ONE OF YOU IS LYING!

"Your eyes can deceive you – don't trust them."

— *OBI WAN KENOBI, STAR WARS: EPISODE IV: A NEW HOPE*

When multiple people witness an event, their stories may not always match, and it's normal for different people to have different versions of events. When details and stories of events don't match, most people blame the differences on human error, imperfect memory, or outright deception. Rarely do people blame the laws of physics.

In Einstein's 1905 paper on Special Relativity, he showed us that **simultaneity**, or events occurring at the same "time," is relative. What appears to happen at the same time for *you* might appear very different to *me*.

To show this principle, he devised another ingenious thought experiment known as the **Train Paradox.** He imagined a train moving from left to right with two observers, one standing perfectly in the middle of the train and one standing stationary outside watching the train. Then, he imagined that just as the observer on the train passed the observer on the ground, two bolts of lightning struck the train, one bolt striking the front of the train and the other bolt striking the back of the train.

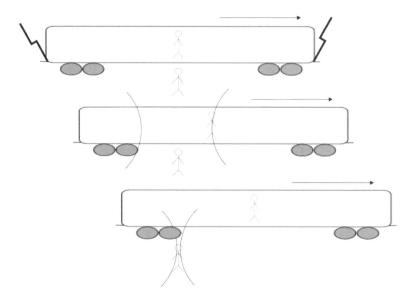

Fig. 6: Train Paradox. Image Credit: Alancrh, licensed under CC By 4.0. Lightning and stick figures added to the original.

What do the observers see?

To the stationary observer, the lightning bolts appear to strike the train simultaneously (at the exact same time). He is outside the train but standing perfectly next to the middle of the train when the bolts strike, so therefore the light from both bolts reaches him simultaneously.

But what does the observer standing on the train see? We know light moves at the same speed in all reference frames, and we also know the observer on the train is moving with the train. And *because* the train is moving, the observer on the train is moving *toward* the lightning bolt that struck the front of the train and *away* from the lightning bolt that struck the back of the train. Therefore, the light from the front lightning bolt reaches the observer *before* the light from the back lightning bolt reaches him, *and therefore he will perceive that the front lightning bolt struck the train first.*

One of them must be wrong, right? Einstein tells us both observers are correct. Simultaneity is relative (Einstein, 18). What happens at the same time to *you* may not happen at the same time to *me*.

7

HONEY, I SHRUNK THE KIDS (BY LAUNCHING THEM AT HALF LIGHTSPEED)

Mad Hatter: "Why is a raven like a writing-desk?"

"Have you guessed the riddle yet?" the Hatter said, turning to Alice again.

"No, I give it up," Alice replied. "What's the answer?"

"I haven't the slightest idea," said the Hatter.

— LEWIS CARROLL, ALICE IN WONDERLAND

So in 1905, Einstein showed us Time is relative – what about Space? Is Space relative too? Yes, says Einstein. Space can bend and stretch too. "Distance" is just what *you* measure between two points, just like "time" is just what *you* measure on *your* clock. Measurements are different for different observers. According to Einstein,

When an object moves, its length contracts in the direction of its movement relative to stationary observers.

In other words, when things go fast, they will shrink to you. The **length contraction** equation, which operates similarly to the time dilation equation in order to keep lightspeed constant to all observers, is shown below:

$$\Delta x = \Delta x' \sqrt{1 - v^2}$$

Eq. 4: Einstein's Length Contraction Equation, where:

$\Delta x'$ = measured length of an object at rest relative to an observer

Δx = contracted measured length of an object in motion relative to an observer

v = speed of an object as a percentage of the speed of light

$\sqrt{1 - v^2}$ = Lorentz Factor

Just like time dilation, this length contraction happens all around us in our everyday lives. When your car moves down the street, for example, its length will contract by a few nanometers – though I doubt your neighbors would notice. The effects only become noticeable at relativistic speeds, speeds approaching that of light (Urone, 2000).

To see length contraction in action, let's walk through another adventure together with Rocket Ralph and Earthly Earl. Let's say Rocket Ralph gets in a space ship that is 100 yards long (measured at rest) and blasts off moving 670 miles per hour (.001% of the speed of light) – what is the length Earthly Earl measures for Ralph's space ship? The solution is provided below:

$$\Delta x = \Delta x' \sqrt{1 - v^2}$$
$$\Delta x = 100yd\sqrt{1 - (0.001)^2}$$
$$\Delta x = 100yd(0.99999999999995)$$
$$\Delta x = \sim 100yd$$

As you can see, the length contraction in this example is much too small to measure, and Earthly Earl will perceive the space ship is still 100 yards long.

But let's make things more interesting.

Let's say that Rocket Ralph increases his speed to a constant velocity of 604,000,000 miles per hour (90% the speed of light). What length does Earthly Earl measure now?

$$\Delta x = \Delta x' \sqrt{1 - v^2}$$
$$\Delta x = 100yd\sqrt{1 - (0.9)^2}$$
$$\Delta x = 100yd(0.435)$$
$$\Delta x = 43.5yd$$

At speeds this high, Earthly Earl observes that the space ship is only 43.5 yards long – less than half its original size. Does Rocket Ralph feel cramped inside his space ship? Not at all, according to Einstein. To Ralph, everything about his space ship is perfectly normal and the size hasn't changed one bit.

Just like time dilation, length contraction works in both directions. To Rocket Ralph, he perceives he is still and Earthly Earl is the one moving past him at 90% the speed of light – therefore he observes that Earl (not him) is the one who is shrinking.

Strange Facts: Careful readers may be wondering a very good question – how much length contraction occurs when an object moves at the speed of light? Unfortunately, particles with mass can never travel at the speed of light (this, we'll see in future chapters, would require infinite energy). Photons on the other hand, which are massless force-carrier particles of light, do indeed travel at the speed of light from the sun to Earth. To us, we measure it takes a photon roughly 8.3 minutes to travel the 93 million miles from the sun to Earth. But to the photon, it is still and the Earth is moving toward it at the speed of light. At this speed, length contraction of the Space between Earth and the sun approaches zero. Therefore, *from the photon's perspective, it arrives to Earth instantly. In fact, from a photon's perspective, every trip is finished instantly, no matter how far away the destination is. To a photon, Time itself is frozen.*

Thus, Einstein concluded his third paper on Special Relativity in June 1905, his Miracle Year. For the next three months, he would continue going to work at the patent office as normal and contemplating his ideas. Then in September 1905, he published a "supplement" to his Special Relativity paper in the same science journal that is called "perhaps the most profound afterthought in the history of science."

He had stumbled on the secret of how stars are born and how the sun shines. He had stumbled on the secret of what would lead to the deadliest weapon mankind has ever produced. He had discovered that mass and energy are different manifestations of the same *thing*.

He had discovered the most famous equation in history.

$E = mc^2$.

PART II

DISCOVERY II

EINSTEIN STANDS ON NEWTON'S SHOULDERS & STUMBLES ON THE WORLD'S MOST FAMOUS EQUATION

8

ON THE SHOULDERS OF GIANTS

"Reality is merely an illusion, albeit a very persistent one."

— ALBERT EINSTEIN

To fully appreciate the genius of Einstein's famous $E = mc^2$ equation, we must first take a quick look at Newton, particularly his accepted ideas about **momentum** and **energy.** In the late 1600s, Newton declared that momentum is the measure of "motion" in an object and is equal to the product of its mass and velocity:

$$P = mv$$

Eq. 5: Newton's Momentum Equation, where:

P = momentum

m = mass

v = velocity

The interesting thing about momentum is that it's a vector quantity, meaning it also has direction. Imagine a semi-truck crashes head-on into a small European smart car while both are moving at the same speed. What do you think will happen? Intuition and experience tells us the vehicles will travel in the direction of the semi-truck because it's bigger.

Now let's make things more interesting. Imagine a 30-ton truck moving 1 mile per hour crashes directly into a 2-ton truck moving 15 miles per hour. Who wins the battle now? Newton says both cars remain perfectly still after the collision because their momentums, 30 units, cancel out perfectly. The big, slow-moving truck has the exact same momentum as the small, fast-moving truck. A small, explosive running back can generate just as much momentum as a 300 lb offensive lineman, Newton says.

Just like with time and space, Einstein says Newton's equation for momentum works well for low speeds, but that when objects move closer to the speed of light their momentum exponentially increases according to this equation:

$$P = \frac{mv}{\sqrt{1-v^2}}$$

Eq. 6: Einstein's Momentum Equation, where:

P = momentum

m = mass

v = velocity as a percentage of light speed

$\sqrt{1-v^2}$ = Lorentz Factor

Next, let's look at Newton's theories on energy, the ability to do work, specifically **kinetic energy**, the energy of an object in motion. Newton believed kinetic energy to be the same as momentum, to be an object's mass times its velocity. His great rival, German scientist Gottfried Leibniz, insisted instead that kinetic energy is equal to the product of an object's mass and the *square* of its velocity.

Strange Facts: This is the same Gottfried Leibniz whom Newton would later accuse of plagiarism for stealing Newton's work and taking credit for the invention of Calculus. Leibniz insisted he had no prior knowledge of Newton's unpublished work, which was later exposed as a lie.

In 1730, Dutch scientist Willem Gravesande put the controversy to rest when he dropped several brass spheres into a clay floor and measured their speed. According to Newton, if a ball moves twice as fast as another ball of the same mass, it will have twice the momentum (and move twice as far into the clay).

But that's not what Gravesande saw. Instead, he found that a ball moving twice as fast as another ball of equal mass would move *four* times as far into the clay. In other words, the ball's momentum is equal to the product of its mass and the *square* of its velocity. Leibniz, not Newton, was correct (Egdall, 91).

Today, we formally recognize that an object's kinetic energy is this:

$$KE = \frac{1}{2}mv^2$$

Eq. 7: Kinetic Energy, where:

KE = kinetic energy measured in Joules

m = mass

v = velocity

In September of 1905, Einstein himself was keenly aware of this equation, and he was about to use it to discover a *new* equation. One that would change the world forever. And he would do it with another bizarre day dream, his **double flashlight thought experiment.**

9

EINSTEIN STUMBLES ON WORLD'S MOST FAMOUS EQUATION

"The truth isn't always a blinding light. Sometimes it's a deep and dazzling darkness, that illuminates – and burns – just as surely."

— ALBERT EINSTEIN

"Hmmm," Einstein scratched his head and drew from his pipe. His office desk, located on the third floor of the Federal Office for Intellectual Property in Bern, Switzerland, was littered with patents and scribbles of mathematics, carefully hidden from his coworkers. It was September of 1905.

"Say I have a flashlight that shines in both directions, resting on a frictionless surface. Would it move?" he asked himself, drawing from his pipe.

"No," he thought. "The momentum of the two light beams would cancel each other out.

"What would the energy of the light beams be if I was resting relative to the flashlight?

"What would the energy of the light beams be if I was moving relative to the flashlight? Would there be a difference?" he wondered.

Then, he wrote down what he already knew to be true, the classic formula for kinetic energy:

$$KE = \frac{1}{2}mv^2$$

He scribbled his energy equations, both from the moving reference frame and from the resting reference frame. He combined the equations and did a little algebra.

The change in kinetic energy of the flashlight was this:

$KE = \frac{1}{2} \left(\frac{E}{c^2}\right) v^2$, where

KE = kinetic energy of the light

c = speed of light

v = relative velocity of the moving frame

Eureka. With trembling fingers, he wrote a new equation.

$$m = \frac{E}{c^2}$$

And finally, he had it:

$E = mc^2$.

In Einstein's short follow up paper to his Special Relativity paper in September of 1905, his double flashlight day dream, inspired by relativity, led him to an extraordinary conclusion, considered by many as the greatest afterthought of all time.

Einstein showed that as the double flashlight was emitting energy , it was also losing an extremely small amount of mass, equal to the energy released divided by the square of the speed of light (Einstein, 34). The flashlight loses weight as it emits light. Thus he concluded:

The mass of a body is the measure of its energy content.

$$E = mc^2$$

Eq. 8: Einstein's Mass-Energy Equation, where:

E = energy

m = mass

c = speed of light

In other words, Einstein was saying that energy and mass are *equivalent*, that they are different manifestations of the same thing. Mass has weight and inertia, and so does energy. *A small amount of mass, therefore, can be converted into a vast amount of energy, proportional to the square of the speed of light.*

SO WHAT?

Equations on a page may seem too abstract for normal life – what does all of this mean for you and me? It means that anything that gives off energy, whether it's a light bulb or a candle or even a toaster, loses a bit of mass too along the way. Consider, for example, burning coal. Before Einstein made history with this mass-energy equation, scientists would say that as the coal burns, its mass is conserved as it turns into ashes and escaping gases. Matter is conserved. At the same time, energy is conserved as the chemical energy inside the coal is converted into heat and light energy. Energy is conserved. Together, these principles are known as the **conservation of energy** and the **conservation of mass**.

Einstein changed this principle by saying that as the coal burns, *a tiny bit of its mass is also converted into radiation*, so that the weight of the ashes and escaping gases that is left after the coal burns is very slightly less than the weight of the coal before it burns. At the same time, the chemical energy of the coal before it burns is very slightly less than the total energy released after it burns because there is a very tiny bit of extra radiation energy that is released along with the light and heat. Thus, Einstein says, energy and mass, considered separately, are *not* conserved. Instead, *mass-energy* is conserved. Today, we recognize this principle as the law of the **conservation of mass-energy** (Egdall, 99).

But Einstein wasn't done. At the end of his paper, he wrote the following words, words he would later wish he had never written:

> *It is not impossible that with bodies whose energy content is variable to a high degree (e.g. radium salts) the theory may be successfully put to the test.*

Einstein was suggesting that scientists pick up the baton and test the mass-energy relationship with a focus on **radioactive** materials, materials with an unstable atomic nucleus which emit electromagnetic radiation, by converting a small amount of their mass into *enormous* energy. In only a few short decades, scientific research in this field would lead to the creation of the most destructive weapon mankind has ever produced – the atom bomb.

But before we uncover the incredible history of the creation of this weapon, we must first pause for a moment to tell the story of another genius. In late 1905, Quantum Physics and Special Relativity were separate branches with no connections between them. But one eccentric man would change that. Just as Maxwell was able to unite electricity and magnetism into one unified theory, this man would unite Einstein's Special Relativity with Quantum Physics in one brilliant stroke. And in the process he would stumble onto the deadliest thing in the universe.

10

A MAD SCIENTIST DISCOVERS THE MISSING LINK (AND THE DEADLIEST THING IN THE UNIVERSE)

"I don't like Quantum Physics because I like to think the moon is there even when I'm not looking at it."

— *ALBERT EINSTEIN*

When Einstein published his first paper in June of 1905 revealing the particle nature of light, he unwittingly birthed an entire new branch of physics, **Quantum Physics**, the incredible physics of subatomic particles. Einstein had shown that light is not just a wave, as Newton had thought, but is actually made up of small quantum **photon** particles. Other physicists would later show that light in fact has **particle-wave duality**, that it can behave like a wave *and* a particle. Over the next several years, the quantum revolution broke loose as the world's top physicists rushed to discover nature's smallest building blocks – the electron, the proton, the neutron, and the quark.

Another eccentric genius, Erwin Schrödinger, sent shockwaves through the world when he showed that it is *impossible* to truly define a particle's location in space and time, and that it was only possible to calculate the *probability* of where a particle was *likely* to be found. In 1926, he derived probabilistic wave functions to describe the motion of subatomic particles moving at slow speeds, and he was instantly recognized for his breakthrough, a version of "Newton's Second Law" for quantum particles. His wave functions revealed one of the most miraculous and maddening principles of quantum mechanics, the principle of **superposition**, the bizarre reality that a particle adopts multiple locations and energy levels at once and only commits to a particular "quantum state" *once it's observed.*

The Man Behind the Math: *Erwin Schrödinger's genius and personal quirks were famous, even in his own lifetime. In a time when such arrangements were considered grossly indecent, he kept two lovers his entire adult life, both his wife and his mistress. Even though he grew up in a Catholic household, he remained a lifelong atheist. He fled Germany in the 1933 because he opposed Nazism and the persecution of Jews but later apologized for the move. He made incredible contributions to Quantum Physics but* *intensely disliked and doubted his own discoveries. Here was a deeply gifted but complicated and troubled man.*

Schrödinger was notorious for doubting his own discoveries and despised his own superposition principle at first because it made zero logical sense. In order to expose the ridiculousness of his own idea, he published his famous "Schrödinger's Cat" thought experiment, in which a cat inside of a box is exposed to a 50% chance of being killed by a deadly poison, and asked, Is the cat alive or dead inside the box? The cat can't be alive and dead at the same time – it must "choose" a state. His thought experiment backfired tremendously, and other physicists in his time, most notably Neils Bohr, used "Schrödinger's Cat" to argue that yes, the cat is alive and dead inside the box – it only chooses one of these two states when someone bothers to peek (Learn, 2021).

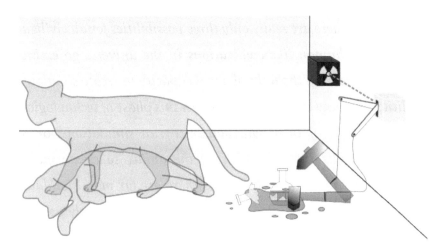

Fig. 7: Schrödinger' Cat thought experiment. Image Credit: Dhatfield, licensed under CC By 4.0.

This experiment, incidentally, inspired the first published theory of parallel universes or multiple realities, the Many Worlds Interpretation. According to this theory, we may observe the cat is dead, but there exists a separate branching reality in which we observe the cat as alive. Each passing moment is another branch point to infinite other realities that never intersect.

How Deep Does the Rabbit Hole Go? *Disturbingly, there is only one other phenomenon in the known universe that mirrors Schrödinger's discovery that particles exist in superposition and are only rendered "real" upon observation – video games. Programmers who are faced with the daunting task of rendering complex video game worlds are unable to do so with limited modern computers. Instead, they make the choice, born of necessity, to only render that which is observed (all unobserved things are unknown). Nick Bostrom, the leader of modern Simulation Theory, says there are really only three possibilities for all civilizations: 1) All human-like civilizations in the universe go extinct before they develop the technological capacity to create simulated realities; 2) if any civilizations do reach this phase of technological maturity, none of them will bother to run simulations; or 3) advanced civilizations would have the ability to create many, many simulations, and that means there are far more simulated worlds than non-simulated ones. Therefore, it is entirely possible that the world we perceive around us is not "base reality," but is* *only an extremely complex simulation nested within countless other simulations. Wake up, Neo.*

The problem with Schrödinger's wave function equations was that they only described the motion of particles at slow, non-relativistic speeds. No one, not even Einstein himself, had yet managed to unite quantum theory with Special Relativity, the behavior of objects moving at speeds approaching that of light.

But in 1931, a young British physicist named Paul Dirac, in a lightning burst of inexplicable inspiration and genius, derived a mathematical equation that broke through the limits of Schrödinger's equation to explain the behavior of an electron moving at near-lightspeeds with spin. *His equation merged all the most important elements of Special Relativity with Quantum Physics in one stroke* (Gerritsma, Kirchmair, Zahringer, Solano & Roos, 2009).

Dirac's electron equation is written as:

Eq. 9: Dirac Equation

$$(\beta mc^2 + c\sum_{n=1}^{3}\alpha_n p_n)\psi(x,t) = \frac{i\hbar \partial \psi(x,t)}{2\pi \partial t}$$

In this equation,

ψ = wave function for the electron with spacetime coordinates x, t

p = momentum

c = speed of light

h = Planck's Constant, $6.62607015 \times 10^{-34}$ *joule − seconds*

Using his equation Dirac was able to describe the momentum, position, spin, and energy of an electron at near lightspeed and their changes over time, one of the greatest breakthroughs in all of modern physics. Dirac's genius, inspired by Schrödinger himself, was to define his wave function, not as a single electron state, but as the Schrödinger-inspired superposition of four separate electron states: a spin-up electron, a spin-down electron, a spin-up positron, and a spin-down positron. Dirac introduced these new positron particles in order to account for a strange result he could not explain – over and over again, his equation yielded solutions that showed negative energy for the electron. How was this possible?

In order to explain his results, Dirac predicted that these negative energies arose from undiscovered particles he named positrons, particles with exactly the same mass as electrons but with positive charge. His equation also predicted that when a positron contacts an electron, the two will **annihilate** each other, with the combined mass transforming into released energy, *exactly according to Einstein's famous equation, $E = mc^2$*. And not only can electrons and positrons turn into energy – the resulting energy can then turn *back* into muons and antimuons, which are similar to electrons and positrons but much more massive. $E = mc^2$ works in both directions. Dirac's peers doubted his predictions until physicist Carl Anderson discovered the positron in lab experiments, and confirmed Dirac's predictions a year later in 1932 (Webb, 2008). Antimatter's ability to destroy matter itself makes it the most dangerous substance in our universe.

Strange Facts: *The problem and danger of annihilation is very real, and modern instruments regularly detect antiprotons in space before they're annihilated. Further, many scientists theorize that the Big Bang should have created matter and antimatter in equal amounts, and that these two classes of matter should have annihilated each other completely, meaning we never should have existed. Their conclusion is that we exist today only because, for every billion matter-antimatter pairs, there was one inexplicable extra matter particle.*

Dirac's work on electron behavior and antimatter, the strange substance that annihilates all matter it comes into contact with, would have been enough to immortalize him forever because he forged a link between Quantum Physics and Special Relativity, but he didn't stop there. In his 1927 paper *The Quantum Theory of the Emission and absorption of*

Radiation, he proposed that photons are not just quantized particles of light- they are "quanta" of the electromagnetic field. From there, he made an astonishing leap and proposed that *all particles*, not just photons, are nothing more than localized vibrations in a quantum field, and that these fields exist everywhere around us (Pratt, 2021).

His **Quantum Field Theory (QFT)** of electromagnetic radiation was another lightning flash of brilliance in the journey toward a Theory of Everything that forced physicists to consider an incredibly new and exciting possibility.

What if every force besides the electromagnetic force, including the strong force, weak force, and gravitational force, is composed of quantized fields?

And what if these forces are carried by particles which are really just vibrations in those quantum fields?

And matter particles, particles with mass?

What if they're just vibrations in quantum fields too?

Paul Dirac would later receive a Nobel Prize in 1933 for his mesmerizing electron equation, which was rightly regarded as the first major step toward a unified Theory of Everything. When asked to comment on his peculiar, introverted friend and his accomplishments, Einstein observed that Dirac "constantly balanced on the dizzying path between genius and madness."

But as these physicists toiled quietly in their offices, changing the world in their own way, another *kind* of madman was slowly readying his armies for what would become the deadliest conflict in human history. His name was Adolf Hitler.

11

THE ROAD TO MANKIND'S MOST DANGEROUS WEAPON

"I know not with what weapons World War III will be fought, but World War IV will be fought with sticks and stones."

— ALBERT EINSTEIN

In August of 1939, Albert Einstein carried the weight of the world on his shoulders. Adolf Hitler, the totalitarian dictator of Nazi Germany, had violated the Munich agreement and invaded Czechoslovakia and was poised to invade Poland as well – peace in Europe and all of western civilization hung in the balance, and although Hitler had struck an agreement with Stalin that he wouldn't invade Poland, Hitler's enemies

THE ROAD TO MANKIND'S MOST DANGEROUS WEAPON | 71

abroad, including Einstein himself, knew better. In a few short months, World War II, the deadliest military conflict in human history, would erupt on the world stage.

Just before the Polish invasion, Einstein observed that Hitler blocked the sale of uranium in Czechoslovakia, one of the great sources of uranium ore in the world, and shrewdly guessed that the Nazis were on the path to develop a nuclear weapon. Such a weapon, he knew, would wield destructive power that would eclipse that of any other weapon ever devised and bring its possessor a decisive advantage to any military conflict. His brilliant guess rested on the spectacular discovery of nuclear fission of uranium atoms by physicists Lise Meitner and Otto Frisch (and its destructive power according to his equation $E = mc^2$) in January 1939, mere months before Hitler occupied Czechoslovakia later that year in March.

Below is the chemical equation for the **nuclear fission** of uranium-235, in which a neutron collides with a uranium-235 atom and causes it to split into lighter elements and three additional neutrons. These neutrons then collide into other uranium atoms, triggering an exponential chain reaction and a *massive* release of energy.

$$^{235}_{92}U + ^{1}_{0}n \rightarrow 3\,^{1}_{0}n + ^{92}_{36}Kr + ^{141}_{56}Ba + \text{ENERGY}$$

Einstein immediately penned a secret letter to U.S. President Roosevelt, writing:

> I understand that Germany has actually stopped the sale of uranium from the Czechoslovakian mines which She has taken over. In the course of the last four months it has been made probable that it may become possible to set up a nuclear chain reaction in a large mass of uranium, by which vast amounts of power and large quantities of new radium-like elements would be generated. This new phenomenon would also lead to the construction of bombs (Groves, 1975).

Einstein then called upon Roosevelt to develop America's own nuclear weapon to oppose the Nazis, and Roosevelt heeded the call, officially placing Robert Oppenheimer as head of the top-secret **Manhattan Project** in 1942. At its height, over 130,000 Americans were working on developing the bomb, spread out across thirty-seven top secret locations in the United States as WWII raged on.

Strange Facts: The Nazis despised Einstein and his theory of Relativity as antisemitism rose in Germany during Nazi rule, so much so that Relativity was dismissed as "Jewish science." Einstein fled to the U.S., becoming a U.S. citizen in 1940, and there were multiple bounties on Einstein's head from Germany, with one magazine labeling his image with the caption, "Not Yet Hanged." Einstein supported the Allies and did far more than just help persuade Roosevelt to build the atom bomb. He wrote to other European leaders like Churchill and Chamberlain urging them to embrace Jewish scientists and employ them in their respective countries. In the process, Einstein helped to save thousands of lives.

By May of 1945, the Allied Forces, including the United States, Russia, and Britain defeated Nazi Germany in Europe and captured Berlin, but war still continued with Japan. Meanwhile, the scientists of the Manhattan Project were just weeks away from developing a fully operational atom bomb. On July 16, 1945, they detonated the first atom bomb in Alamogordo, New Mexico at a military test facility. The explosion erupted in a brilliant flash of light and radiation, followed by a giant mushroom cloud over the desert floor. Windows of houses more than fifty miles away shattered in the wake of the catastrophic nuclear blast - the nuclear era had begun (Metcalfe, 2023).

News of the successful detonation of a nuclear weapon reached Truman quickly, but half a world away the Allies still battled the Japanese, who staunchly refused to surrender. Truman's military advisers calculated that a land invasion of Japan would effectively end the war but would come at a heavy price, costing tens or hundreds of thousands of lives of American as well as Japanese soldiers and civilians. They presented the nuclear weapon as an alternative to a costly ground war, Truman accepted, and after Japan refused to respond to Truman's threats of "prompt and utter destruction," Truman authorized the use of the atom bomb on Japan.

On August 6, 1945, an American B-29 bomber dropped the atom bomb on Hiroshima, Japan, devastating the city in an instant and killing 140,000 Japanese civilians. Still the Japanese government refused to surrender. Three days later, on August 9, an American bomber dropped a second nuclear weapon over the city of Nagasaki, Japan, killing over 80,000 Japanese civilians. On August 15, the Soviet Union officially declared war on Japan, and Japan finally issued unconditional surrender, bringing the war to an end.

The atom bomb helped to usher in a lasting era of peace, but not without consequences. Spies in the Manhattan Project, most notably Klaus Fuchs, handed nuclear secrets to the Soviets, and by 1949 the Soviet Union had devised its own nuclear weapon, and a new kind of conflict, the Cold War, spanned the next five decades until the Soviet Union finally collapsed economically in 1991. Nuclear weapons haven't been used in military conflict since 1945 (Kelly, 2005). Still, the threat of nuclear war remains the ultimate safeguard of peace between the major powers of the world to this day.

Einstein himself, a passionate pacifist, did not appreciate the "peace" he had helped to usher in through the creation of the atom bomb. Instead, the hundreds of thousands of lives lost weighed heavily on him till his death. The hindsight knowledge that Germany never successfully constructed a bomb themselves drove him sick with regret, and he wished he had never signed his letter to President Roosevelt.

"Had I known that the Germans would not succeed in producing an atomic bomb," he said, "I never would have lifted a finger."

Congratulations, you're well on your way to finishing this book. If you're gaining value from the content so far and would like us to publish future books, please let us know by leaving a 5 star review. We appreciate it! Scan the QR code below to leave your review:

PART III

DISCOVERY III

EINSTEIN DAY DREAMS ABOUT FALLING & SHOCKS THE WORLD WITH A NEW THEORY OF GRAVITY

12

EINSTEIN'S OLD PROFESSOR STUMBLES ONTO SPACETIME

"[Einstein's 1905 paper on Special Relativity] came as a tremendous surprise, for in his student days Einstein had been a lazy dog."

— HERMANN MINKOWSKI

"You've got to be kidding me. Albert wrote this?"

Einstein's old college mathematics professor from Zurich Polytechnical Institute, the great mathematician Hermann Minkowski, devoured Einstein's articles on Special Relativity in the Annalen der Physik journal in a single sitting with breathless excitement.

"This kid took nine math courses under me and spent more time skipping and smoking his pipe than he did in my classroom," Minkowski muttered to himself.

"And yet here he is, defying Newton and saying that Time and Space are relative? This slacker, this amateur? This boy? Unbelievable."

Minkowski took some scratch paper out of his pocket and started jotting down his thoughts.

"So Albert is saying that as we move through Space, Time slows down for us relative to observers. And the faster we move through Space, the more time dilation occurs, the more Time slows down. And Albert is also saying that distance contracts in the direction of travel. Therefore, the space interval and the time interval of the journey as measured by different observers are not only relative – they are connected. But how?"

Three years later Minkowski delivered a talk entitled "Space and Time" in Cologne, Germany to physicists from all over the country and proposed that Space and Time are in fact connected in what he dubbed "four-dimensional spacetime."

Einstein would shrug off the idea at first, criticizing his former professor for mathematical acrobatics and saying, "With math you can prove anything."

In time, however, Einstein would grow to appreciate the mathematical genius of his professor. Little did he know that he would *someday **need** Minkowski's equations of spacetime in order to complete his life's work, a new theory of gravity – his theory of General Relativity.*

How did Albert Einstein's old math teacher make this discovery? Minkowski's first insight was to imagine a coordinate system for spacetime, with the horizontal axis representing space and the vertical axis representing time. Every event in the universe has a specific coordinate on his graph, a specific point in Space and a specific point in Time. According to Minkowski, everything in the universe has its own "clock" attached to it, and as it moves through Space or simply continues to exist (moving forward in Time), it plots its own **worldline** through spacetime.

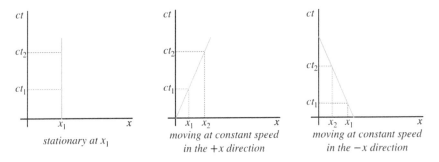

Fig. 8: Minkowski Spacetime Diagrams for an object at rest and for an object in motion. Space is the x axis while Time is the y axis. Image Credit: Tom Weideman, licensed under CC By 4.0.

But Minkowski wasn't finished yet. His next step, his brilliant breakthrough, was to look at Einstein's Lorentz transform again and reach a conclusion Einstein had failed to see - he proved for spacetime what Pythagoras had proven for right triangles. Imagine two random events occurring, such as an apple falling out of a tree and the same apple landing on the ground. As Pythagoras had shown that the square of the hypotenuse in a right triangle is the sum of the squares of the other legs, *Minkowski proved that for any two events occurring in the universe, the square of the spacetime interval between them is equal to the difference of the square of the time interval between them and the square of the space interval between them. Even more incredibly, this spacetime interval is absolute and invariant (doesn't change) with relative uniform motion - it is the same for all observers. The space interval between two events is relative. The time interval between two events is relative. But the spacetime interval for two events is absolute – it is always the same.* Below is his famous spacetime formula:

$$(\text{Spacetime Interval})^2 = (\text{Time Interval})^2 - (\text{Space Interval})^2$$

$$\Delta S^2 = \Delta t^2 - \Delta x^2$$

Eq. 10: Minkowski's timelike spacetime formula, where:

ΔS = spacetime interval

Δt = time interval between two events

Δx = space interval

Just as Einstein had shown that the speed of light is absolute and invariant with relative uniform motion, his old professor later showed that the spacetime interval is absolute as well (Egdall, 119). So what does this mean exactly? Let's walk through a simple example to appreciate how bizarre this discovery really is.

A deadly meteor is hurtling toward Earth. To save the planet, NASA launches a nuclear warhead from the surface (Event 1) which then hits the meteor (Event 2), causing it to splinter into several pieces that harmlessly pass by the Earth. NASA mission control, the resting frame, measures it takes the warhead 5 seconds to travel 4 light-seconds (about 745,000 miles) and impact the meteor. At the same time, a spaceship travelling 25% of the speed of light at constant speed is passing Earth and observing the same two events. Due to time dilation and length contraction, the spaceship measures the warhead takes 4.13 seconds to travel 2.84 light-seconds to impact the meteor.

> a. What is the spacetime interval for these two events according to NASA, the resting frame?

$$\Delta S^2 = \Delta t^2 - \Delta x^2$$
$$\Delta S^2 = (5s)^2 - (4\ light-seconds)^2$$
$$\Delta S^2 = 9s^2$$
$$\Delta S = 3s$$

b. What is the spacetime interval for these two events according to the spaceship, the moving frame?

$$\Delta S^2 = \Delta t^2 - \Delta x^2$$
$$\Delta S^2 = (4.13s)^2 - (2.84\ light-seconds)^2$$
$$\Delta S^2 = 9s^2$$
$$\Delta S = 3s$$

Both NASA and the spaceship observe the same thing and observe different time and space measurements, but the spacetime interval for both of them is the exact same – 3 seconds.

The fact that the spacetime interval doesn't change for different observers struck Einstein initially as a mathematical trick of little importance, but it would have far-reaching philosophical implications for his later work on General Relativity. In a world of constant motion, *here* finally was an absolute method to measure events in Space and Time to restore order to the chaos. Here was mathematical proof that Space and Time are connected (Egdall, 121).

By 1905, Einstein had fundamentally shown that Newton was wrong about light, Time, and Space. His next task would prove far more ambitious, and far more difficult. Einstein was about to question one of the most widely proven and successful theories in all of science – Newton's theory of gravity.

13

PROBLEMS WITH NEWTON'S GRAVITY

"As an older friend, I must advise you against [challenging Newton's gravitation theory] for in the first place you will never succeed, and even if you succeed, no one will believe you."

— MAX PLANCK, IN A LETTER TO EINSTEIN

In the early 20th century, absolutely no one in the physics community dared to question Newton's gravity theory. *Why fix what isn't broken?* For over 200 years, Newton's law of gravitation, published in his masterpiece *Principia* in 1687, had worked beautifully and been confirmed through observation and experiment dozens and dozens of times.

Eq. 11: Newton's Law of Gravitation.

$$F = G\frac{m_1 m_2}{r^2}$$

In this equation,

F = gravitational force between two objects

G = gravitational constant

r = distance between the objects

$m_1 m_2$ = masses of separate objects

Newton's equation perfectly described the motion of the planets around the sun, the motion of the moon around the Earth, the tides, and it even successfully predicted the existence of Neptune and Pluto *before* they were ever observed with telescopes. It was considered one of the great triumphs of science.

Strange Facts: *Isaac Newton, though regarded in his time as a modern scientist, was bizarre, even for a physicist. When the bubonic plague seized London in the 1660s, he published an article describing what he believed to be the cure – toad vomit. He was also a closet Medieval wizard who was secretly obsessed with alchemy, the study of turning base metals into gold. He spent 30* *years trying to create gold in his office – what other contributions would he have made to science and mathematics if someone had snapped him out of this obsession?*

There was only *one* problem with Newton's gravity theory in the early 20th century – Newton's equations did not perfectly fit Mercury's orbit. Instead of questioning Newton, most physicists either ignored the problem completely or theorized instead that perhaps a small, undiscovered planet between the Sun and Mercury was throwing off Mercury's orbit slightly. So small was this flaw that it seemed trivial in comparison to the long list of successful predictions and measurements flowing from Newton's gravitation theory (Inglis-Arkell, 2013).

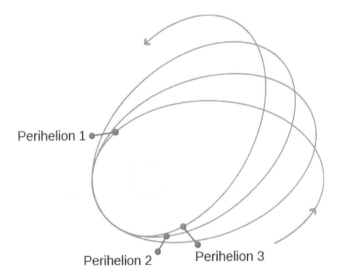

Fig. 9: Mercury has a perihelion orbit that shifts precisely 42.9 arc seconds per century. Image Credit: University of Florida Astronomy Department, licensed under CC By 4.0.

But when Einstein published his Special Relativity theory in 1905, he was forced to confront three additional problems with Newton's gravity theory because they *directly* clashed with his own ideas. In order to properly defeat the legendary Newton in intellectual battle, Einstein would be forced not only to expose these problems, but to provide an alternative theory that solved them *(and matched all of Newton's prior successful predictions and observations as well)*. This task would prove tougher and more difficult than anything he had previously accomplished – indeed it would become his sole obsession for the next decade of his life.

First, Newton believed that gravity is instantaneous, which violated Einstein's idea that nothing in the universe can travel faster than light. If, for example, the sun magically disappeared, Newton predicted the Earth would *immediately* fly off into space, free of the sun's gravity. Einstein, however, predicted that if the sun disappeared, the information would travel at the speed of light to Earth and take just over 8 minutes to arrive. Only then would the Earth fly off into space.

Second, Newton's gravitational law says the gravitational force between two objects decreases as the square of the distance between them – but what exactly is "distance"? Einstein says that the distance between any two points in space contracts with relative uniform motion in the direction of that motion. So the distance between two points as measured from one point is never perfectly the same as the distance measured from the other point – distance is different for different observers. Newton's equation failed to account for this.

Third and finally, Newton was never able to explain *how* the gravitational force acts over distance. His three laws of motion dealt with objects in contact, but gravity acts over great distances – what was the mystery of how it worked? Was it a wave? A particle? Something else? Newton himself actually admitted ignorance to these questions in *Principia*, saying, "I have not been able to deduce the cause of these properties of gravity and offer no hypothesis" (Egdall, 146).

And so Newton glossed over this glaring, inconvenient mystery, and for the next two hundred years no one, particularly the academic elite, dared or bothered to challenge Newton's theory of gravity or even reexamine the issue. That is, of course, until a little-known, rebellious patent clerk decided to risk everything to challenge Newton's ghost.

14

EINSTEIN DAY DREAMS ABOUT FALLING, HAS THE HAPPIEST THOUGHT OF HIS LIFE

"Imagination is more important than knowledge."

— ALBERT EINSTEIN

The year was 1907. Einstein sat at his desk at the patent office, deep in thought about the mystery of gravity. Particularly, he was intrigued by Galileo's assertion that all bodies accelerate to the ground at the exact same rate, regardless of their mass.

"Say Albert, how is the triode patent coming along?" his supervisor asked as he walked by Einstein's desk.

Einstein nearly fell out of his chair with surprise.

"Uh, Sir, it's coming along nicely." Einstein slid his physics scribbles out of view. "There's a chance the idea may not be entirely original. I'll have my full analysis on your desk by Friday," he told his supervisor.

"Thanks, Albert. Stay the course – you've got a promising career ahead."

Einstein sighed with relief as he watched his supervisor walk away.

At that moment, he was struck with a thought he would later describe as the "happiest thought of his life."

When his supervisor surprised him and he nearly fell out of his chair, for a moment as gravity seized him Einstein felt weightless.

Was this weightlessness not the exact same feeling one would feel away from the gravitational pull of Earth, in the middle of outer space?

In fact, Einstein thought to himself, if he was in free fall, there would be no way to tell he wasn't actually in outer space except for the sight of the Earth accelerating toward him very, very quickly. In free fall, gravity would be imperceptible (neglecting air resistance) – instead, it would appear to him as though he were perfectly still and the Earth was accelerating toward him instead and not the other way around.

Einstein had stumbled on one of the most revolutionary discoveries of all time.

Gravity and acceleration, Einstein realized, are equivalent.

Einstein illustrated this principle, again not with fancy mathematics, but with his ingenious **space elevator thought experiment.** Imagine two elevators, one on Earth and one in outer space. In the Earth elevator, imagine a man drops a bowling ball and a golf ball. According to Galileo and Newton's observation that all things in Earth's gravitational field cause objects to fall at the same rate g, the bowling ball and golf ball hit the ground at the exact same moment.

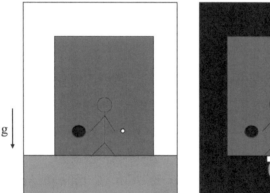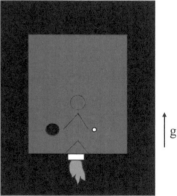

Fig. 10: Einstein's space elevator thought experiment. A man in an elevator under a gravitational force g will experience the same physical effects as a man accelerating upward in an elevator in zero gravity outer space with acceleration g. Image Credit: Author's work.

Now imagine a second elevator suspended in outer space with zero gravity. This second elevator has a small rocket booster attached to the bottom of it causing it to move "upward" through space at the exact same acceleration as Earth's gravity g. An astronaut is floating outside the space elevator recording its movement.

What does the man inside the space elevator experience? He drops the bowling ball and the golf ball. Both are stationary in the zero gravity – they do not "fall." Instead, the elevator itself is moving upward with g acceleration, and the floor of the elevator rises to meet the bowling ball and the golf ball, colliding with them at the same time. *The astronaut outside the elevator sees that the bowling ball and golf ball are still while the space elevator is accelerating upward, but to the man inside the space elevator, the bowling ball and the golf ball have fallen while he has remained completely still.*

Formally, Einstein's great insight of the equivalent effects of gravity and acceleration is known as the **equivalence principle**.

Suddenly, he now had a theoretical basis to declare that acceleration and gravity are relative, because what one observer may observe as gravity, another observer may observe as acceleration.

And with this brilliant flash of insight, Einstein was able to expand his relativity postulate outside of just uniform, constant motion and intuit the following:

All laws of physics are the same for all uniformly moving reference frames, accelerating reference frames, and gravitational reference frames.

In other words, the laws of physics are the same in every situation, whether you're moving at a constant velocity, changing directions, speeding up, slowing down, or falling into the gravitational field of Earth (or any other planet or celestial object for that matter) (Pössel, 2005).

But Einstein wasn't finished yet. First, he recalled from his theory of Special Relativity that uniform motion through space causes time and space to contract and change relative to different observers.

Then, he made a very simple connection:

If uniform motion (constant velocity) causes space and time to warp, why wouldn't acceleration warp space and time too, since acceleration is just changing velocity?

And finally, he applied his equivalence principle:

And if gravity and acceleration produce the same physical effects, then gravity itself must also warp space and time too.

But how?

15

SHOW ME THE LIGHT

"You understand, what I need to know is exactly what happens to the passengers in an elevator when it falls into emptiness."

— ALBERT EINSTEIN TO MADAME CURIE

By 1907, Einstein was stitching together the pieces of what would become his greatest work, his new theory of gravity, General Relativity. But long before he had the mathematical ability to prove his theory correct (he would need outside help for this), he was forced instead to rely solely on the powers of his own imagination, working all by himself with absolutely zero assistance except from his wife Mileva, who lovingly proofread his papers and acted as a sounding board for his early ideas.

Fig. 11: Einstein and Mileva Maric. Author unknown, licensed under CC By 4.0.

Strange Facts: *Mileva Maric and Einstein were both passionate physicists who admired and respected each other greatly. Unfortunately, their marriage would slowly fall apart over time as Einstein's star continued to rise and he began to walk in the circles of Europe's intellectual elite, relegating his former collaborator and lover to the role of dutiful housewife and mother of their two young boys. In 1912, Einstein began a romantic correspondence with his first cousin, Elsa Löwenthal. In 1919, Einstein finally divorced Mileva and married Elsa, promising Mileva a portion of his Nobel Prize money as part of the divorce settlement – he hadn't* even won the prize yet. Incredibly, Einstein did finally win the Nobel Prize in 1921 for the photoelectric effect – the leading physicists of the day were not yet ready to recognize him for his work on Special and General Relativity because it was unproven and too far ahead of its time.

Next, Einstein was curious to know: *how does gravity affect light?*

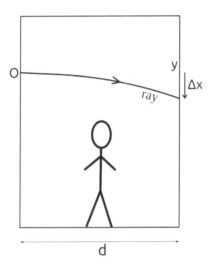

Fig. 12: Elevator light thought experiment. Image Credit: Mathieu Rouaud, licensed under CC By 4.0.

To answer this question, he invented yet another bizarre thought experiment, his **elevator light thought experiment.** He imagined an elevator perched high above the ground with the ability to move in perfect free fall down its cable. Inside the elevator he imagined a small laser on one wall of the elevator firing a single photon at a detector placed on the opposite wall, with both the laser and detector placed at equal height. Next he imagined two separate observers, one standing inside the elevator and one standing on the ground below, watching the elevator drop in perfect free fall.

On signal, as the elevator is in free fall, the laser fires the photon. *What do the observers see?*

To the person inside the elevator, everything inside the elevator is relatively still to her, and she watches the photon travel in a perfectly straight line from the laser to the detector. But the person standing on the ground watching the elevator sees something very different. He sees the laser fire the photon, and as the photon travels to the detector, he sees the elevator continue to accelerate downward so that the detector is in a lower position when the photon finally hits it. Therefore, he observes that the photon follows a slightly *curved* path from the laser to the detector. And since the equivalence principle tells us gravity and acceleration have the same effects, Einstein concluded:

A gravitational field causes light to bend.

But, as usual, Einstein wasn't quite finished yet. He was about to show how gravity warps time as well, and he would do it, as usual, with the power of his own imagination.

Next, he wondered, *how does acceleration (and gravity) affect the frequency of light?* This time, he imagined the same elevator in free fall, but this time he imagined a laser placed on the ceiling of the elevator firing a light beam downward straight through the floor. What would his two observers see this time?

Einstein imagined the person inside would measure the frequency of the light beam as normal, since the everything inside the free-falling elevator is perfectly still to her. Nothing strange to report.

But what would the other observer standing on the ground see? To him, the elevator is racing downward, and so is the laser source. To him, the **Doppler Effect**, discovered over fifty years before in 1847, is causing the frequency the frequency of the light to appear higher than before (just like the frequency would appear lower if the source was moving away). Then, Einstein applied his equivalence principle to the accelerating elevator to conclude:

A gravitational field affects the frequency of light.

And because we learned in earlier chapters that as the frequency of light increases, its energy increases as well, we can also say the following:

A gravitational field affects the energy of light.

This effect is observed even today. When light from distant stars arrives to Earth, it is affected by Earth's gravity and appears to have a higher frequency (a blue-shift). At the same time, when Earth sends light or any other electromagnetic radiation into the atmosphere, the light loses energy and appears to have a lower frequency (a red-shift) as it climbs out of Earth's gravitational pull.

Finally, Einstein made one more stunning leap of brilliance in his thinking about his elevator light experiment. Recalling that **frequency** is nothing more than a measure of wave cycles passing by per unit of time, Einstein realized that for the person inside of the falling elevator, the frequency of the light (and time itself) is behaving normally. *But for the person on the ground watching the elevator and laser source accelerate downward toward him, he would observe that the frequency of the light inside the elevator is higher, and therefore he would see that time is passing faster inside the elevator than it is passing on the ground, according to his wristwatch.* And the accelerating elevator has the same effects as gravity per the equivalence principle. Therefore, Einstein made his final jaw-dropping conclusion:

Gravity affects the passage of time.

Over a half century later in 1971, scientists Joseph Hafele and Richard Keating put Einstein's ideas to the test by running a simple experiment – they placed one highly accurate atomic clock at rest on the ground and a separate atomic clock on board an airplane which traveled around the globe. If Einstein's Special Relativity was correct, the clock on the moving plane would tick more slowly than the clock at rest on the ground, but this slowing would be outweighed by the fact that according to General Relativity, the clock on the airplane would tick faster than the clock on the surface due to weaker gravitational effects. (Gravity is stronger at the Earth's surface and causes time to pass more slowly) The time dilation results matched Einstein's predictions to within 10%.

Strange Facts: *Time dilation occurs all around us in daily life near all objects with mass, even on extremely minor scales that are difficult to measure. For example, you would experience time more slowly at your house than you would if you stood at the top of a very high skyscraper. Similarly, if you were orbiting Earth, time would progress faster than if you were on the surface, which explains why clocks on GPS satellites orbiting Earth regularly fall out of sync with clocks on the Earth's surface and must regularly be re-adjusted. Note that large masses and high speeds can alter the rate of time's passage but that it always moves forward – as of today, backward time travel remains theoretically impossible (unless you happen to own a Flux Capacitor).*

In November of 1907, Einstein quietly published his ideas about how gravity bends light and warps time as an "update" to his Special Relativity theory in *Yearbook of Radioactivity and Electronics*. Slowly but surely, his bold ideas spread to the established communities of physicists in the European universities, the elite circles who had coldly regarded him as an outsider initially (Egdall, 169).

Who was this man?

The father of Quantum Physics himself, Max Planck, read Einstein's work on Special Relativity and immediately saw something in Einstein which few others could – greatness. He wrote, "Einstein's relativity probably exceeds in audacity everything that has been achieved so far in speculative science."

Things began to move quickly for the young patent clerk. In 1909, he received a job offer as Associate Professor at the University of Zurich. Finally, the universities were calling on him with open arms. After some discussion with his wife, Mileva, he accepted the position and left his job as a patent clerk (much to the dismay of his supervisor).

Later that year in October, Einstein received his first nomination for the Nobel Prize – but not for relativity, which he had been unable to conclusively prove yet. Instead, he received the nomination for his 1905 paper on the photoelectric effect, which showed the particle nature of light and was a crucial first step in establishing the most exciting new field of physics in that time, Quantum Physics.

Universities began to fight over Einstein. In April of 1910, the German University in Prague offered Einstein a position as full professor at *twice* the salary he enjoyed in Zurich. Max Planck himself wrote his recommendation. So badly did the German University want Einstein that they would grant him the position of "professor ordinarius," a tenured position in which he would be given the freedom to give up his teaching responsibilities and focus solely on his own work.

As his celebrity rose in Prague, unfortunately, his marriage with Mileva began to falter. Gone were the days of intimate collaboration – instead, Mileva was forced to adopt the role of hostess and housewife, entertaining the endless procession of Einstein's physicist friends and colleagues at their home but never being allowed to participate herself.

Near the end of 1911, Einstein received a call from his old friend Marcel Grossmann at Zurich Poly, the very same university which had refused to offer Einstein a single teaching position after graduation twelve years earlier – Grossmann wanted Einstein to return to his alma mater to accept a full professorship. How sweet and strange the feeling must have been, to be courted by the same university which had rejected him so many times before. Ready to return to familiar friends and eager to save his marriage, Einstein accepted the offer and returned to Zurich.

16

GRAVITY WARPS SPACE

"So, now I too am an official member of the guild of whores."

— ALBERT EINSTEIN ON BECOMING A PROFESSOR

The year was 1912. It had been five years since Einstein showed that gravity warps time and seven years since he published his Special Relativity theory. Einstein, no longer an obscure patent clerk, was a tenured professor at the Zurich Polytechnic Institute, the official Professor of Theoretical Physics. His brave revolutionary work, now famous, had made him one of the brightest stars in physics in Europe, indeed the world.

"The climb did not come easily," Einstein thought to himself as he drew from his pipe.

There had been many critics of his theory, several talented physicists looking to poke holes in his work.

Einstein turned his mind to the most interesting challenge he'd seen yet, a paradox posed by his friend Paul Ehrenfest three years earlier in 1909. Ehrenfest imagined a spinning disc. According to basic Euclidean geometry, the circumference of the disk, Ehrenfest says, is equal to 2pi multiplied by the radius of the disc. Simple enough. But if the disc is spinning, Ehrenfest wondered, wouldn't Einstein's Special Relativity require that the circumference of the disc contract in the direction of its movement (the radius, which is perpendicular to the direction of movement, would be unaffected by length contraction)?

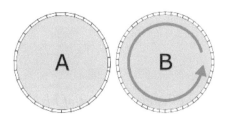

Fig. 13: Spinning Disc. According to Special Relativity, the motion of a spinning disc B will cause length contraction of its circumference in the direction of its motion, causing its circumference to be smaller than the circumference of the identical disc A at rest. Image Credit: Prokaryotic Caspase Homolog, licensed under CC By 4.0.

Therefore, Enrenfest insists, if Special Relativity is to be taken seriously, then the circumference of the spinning disc would actually be slightly smaller than the product of 2pi and its radius, as measured by an observer watching the spinning disc from above.

In other words, Ehrenfest says, the spinning disc would warp space.

Einstein contemplated the spinning disc one more time with fresh eyes.

"The disc is rotating at a constant velocity in the **Ehrenfest Paradox**," Einstein thought to himself.

"But the rotation of the disc means that it is constantly changing directions.

"This means the disc is accelerating, since it is changing directions.

"And we know that gravity and acceleration cause the same physical effects.

"Eureka."

In a flash of insight, Einstein rose out of his chair with breathless excitement and concluded the following:

> According to the equivalence principle, space is warped in a gravitational field (Einstein, 56).

By 1912, Einstein had the basic ideas in place for a new gravitational theory – gravity warps space and time (Egdall, 170). Unfortunately, he didn't have the training or ability to *prove* it mathematically – not yet. He realized that in order to properly show how gravity warps space and time, he would need a new *kind* of geometry that could stretch beyond the limits of Euclidean geometry, one that could accommodate the bending and curvature of space according to his Special Relativity.

As fate would have it, the very same friend who invited Einstein to teach at Zurich Poly in the first place, Marcel Grossman, was a renowned mathematician, chairman of the mathematics department, and he had published seven papers and an entire dissertation focused on a new, cutting-edge field in 20th century mathematics.

Non-Euclidean geometry.

17

EINSTEIN'S OLD FRIEND SHOWS HIM THE WAY

"Something very great will happen to your son someday."

— MARCEL GROSSMAN TO ALBERT EINSTEIN'S MOTHER

"Grossman, you've got to help me or I shall go crazy."

— ALBERT EINSTEIN

After years of continual hitmaking success in physics, Einstein had finally trudged into a depressing mental drought. Ironically, his outward fame and career prestige had never been greater – inside, however, he was completely distraught.

The trouble was that his Special Relativity theory was *limited*. It only applies to reference frames moving at constant velocity in the same direction (ie., non-accelerating reference frames). This, Einstein was painfully aware, is nothing like the real world, where everything in the cosmos is in a state of perpetual acceleration. His Special Relativity theory was not equipped to describe accelerating reference frames, and by extension it was not equipped to describe gravity.

He did reserve a ray of hope, however, in the work of his old professor, Minkowski, on spacetime. Einstein recalled Minkowski's old mathematical discovery that the *spacetime interval is invariant (doesn't change) in uniformly moving (non-accelerating) reference frames*. Thus, Einstein realized that the spacetime interval *does* change moment by moment in non-uniform (*accelerating*) reference frames. And here we see his equivalence principle at work again in 1912:

> *The spacetime interval changes in a gravitational field.*

Einstein realized that *locally*, over a small enough time interval and small enough space interval, the acceleration of a moving object or reference frame would be imperceptible and thus the spacetime interval would be absolute or "flat" – here his Special Relativity theory worked because for this infinitely small interval of space and time he could pretend the object was in uniform motion. But *globally*, Einstein realized, acceleration (and therefore gravity) would cause the spacetime interval to warp and change and *curve*.

To understand the nature of local flatness and global curvature in a tangible way, imagine the surface of the Earth. "Locally" you may say that the ground is perfectly flat, but "globally" over a far enough distance, if you observed the Earth from space you would see the Earth's surface is actually *curved* and spherical in shape. It is only by "stitching together" a vast number of *local* points that you are able to appreciate *global* curvature. Locally flat, globally curved.

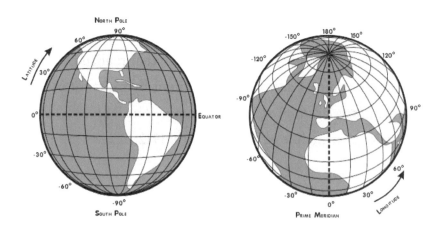

Fig. 14: Locally flat, globally curved Earth. Image Credit: Djexplo, licensed under CC By 4.0.

In 1912, the Great Challenge before Einstein was to mathematically describe this curvature of spacetime in a gravitational field in a way that accomplished eight ambitious goals (among many, many more):

1. His equations needed to match Newton's successful results perfectly. His new gravity theory would not survive if his equations failed to explain how mass (or mass-energy) causes gravity, or if they failed to explain the motion of objects moving at slow, non-relativistic speeds or the motion of the planets and the moon as Newton had done.
2. As a bonus, his equations needed to fully explain Mercury's mysterious orbit, which followed a strange perihelion path that shifted at a precise rate of 42.9 arc seconds per century (Einstein, 73). No one could explain this mystery yet.
3. His equations needed to explain how gravity "acts at a distance" – Newton himself was unable to explain this and didn't even bother to try.
4. He needed to prove that gravity isn't instantaneous (as Newton believed) and that it actually travels at the maximum universal speed limit, the speed of light, in accordance with his own Light Postulate.
5. His equations needed to take length contraction and the relativity of distance into account, something Newton neglected (Einstein, 71).

6. His equations needed to respect and obey his own law of the conservation of mass-energy, which he derived from his spectacular September 1905 discovery that mass and energy are interchangeable, that $E = mc^2$.
7. Five years earlier, he had boldly expanded Galileo's Dictum to say that *all laws of physics are the same for all reference frames (uniformly moving reference frames, accelerating reference frames, and gravitational reference frames)*. Now, he needed to mathematically *prove* this assertion to satisfy the **principle of covariance** – that his equations truly worked in all reference frames.
8. In a classic Cartesian coordinate system, the space and time intervals in Minkowski's spacetime diagrams for Special Relativity are equally spaced apart (in other words the spacetime interval is absolute and doesn't change). But since Einstein knew the spacetime interval dynamically changes in a gravitational field, Einstein was forced to throw this coordinate system out the window. He would need to construct a new *continuously changing* spacetime coordinate system in which the *background* of the coordinate system itself (positions in spacetime) could shift and change, just like spacetime changes in the real world (Egdall, 198).

There is little wonder why Einstein, the talented day dreamer who notoriously shunned his college math classes, struggled in his journey. In the end, he would accomplish the task through gritty, painstaking trial and error in a battle that would span nearly a decade. And he would do it with his old friend by his side, Marcel Grossman.

When Einstein approached his friend and asked for advice on how to map four-dimensional spacetime with a dynamic, changing coordinate system, Marcell replied, "Euclid can't help you. Look to the greats."

"Who's that?" Einstein asked.

"Gauss and Reimann," Marcell replied.

DISCUSSIONS OF POTATOES AND FISHNETS

In the early 1840s, mathematical genius and child prodigy Johann Carl Friedrich Gauss, the founder of non-Euclidean geometry, was tasked with surveying the Kingdom of Hannover in three dimensions, hills, valleys, bumps, and all. To do this, he developed a "curvilinear" (rather than straight-line) coordinate system and applied a network of what are now known as **Gaussian coordinates** directly to the rough, hilly, curved land surface itself.

To properly appreciate the significance of this, imagine throwing a fishnet over a potato. Each little segment of the fishnet bends and stretches and strains over the surface of the potato so that instead of having several neat, equally spaced parallelograms of fishnet, you see several "warped" parallelograms. Now imagine you kept applying fishnets with smaller and smaller and finer and finer line spacings – eventually you would end up with a potato covered with countless fishnet parallelograms that were *locally flat*. Locally flat, globally curved.

Fig. 15: A Fishnet On A Potato – A Continuously Changing Coordinate System for a Curved Surface. Image Credit: Author's work.

Gauss's next step was to describe each parallelogram segment with three "g's" (Gaussian metrical coefficients): length, width, and interior angle. Armed with a list of the three "g's" for every parallelogram segment, Gauss realized he now had the information he needed to mathematically "stitch together" all the locally flat segments to construct a global map of the potato (or any curved surface). Locally flat, globally curved.

Gauss had developed a system to map curved surfaces with a coordinate system that could dynamically change – it didn't matter if Gauss threw his fishnet over the potato at a different angle or rotated the net or shifted its position. He could *still* measure the three "g's" of every parallelogram segment and map the potato successfully. In other words, any reference frame would do because his coordinate system was generally covariant – *this is exactly what Einstein was looking for.*

Ten years later in 1854, Gauss's greatest student, Georg Friedrich Bernhard Riemann, developed a method to extend Gauss's 2-D geometry to any number of dimensions (3-D, 4-D, 5-D, etc.). Riemannian geometry was immediately recognized as "one of the greatest masterpieces of mathematical creation and exposition." Einstein now had a method to apply Gauss's non-Euclidean geometry to four-dimensional spacetime (Egdall, 205).

In addition to Gauss and Riemann, Grossman introduced Einstein to the work of Italian mathematician Gregorio Ricci-Curbastro. *Ricci specifically had already developed surface curvature mathematics based on the Reimann tensor which was also covariant – in other words, Grossman believed the Ricci Curvature Tensor was perfect for mapping the global change in the spacetime interval due to the presence of mass-energy (ie, mapping 4-D spacetime).*

Armed with the insights of these great mathematicians and with Grossman at his side, Einstein developed his very first field equations for General Relativity. The problem? Unfortunately, he found that when he plugged in a specific mass-energy, his equations yielded multiple equally valid solutions for the gravitational field, a sure sign that something was wrong because he needed his field equations to produce a single, correct answer for each mass-energy input.

Stricken with grief, he abandoned the mathematical approach and tried other methods over the next three years, all of them unsuccessful. By now a celebrity professor in Berlin, the world capital of physics, Einstein made a naïve error in *sharing* his intellectual woes with several physicists in public talks – now, he had competitors hot on his trail, racing to find the correct field equations first (most notably a German mathematical genius named David Hilbert). To make matters worse, his marriage with Mileva was falling apart beyond repair – she left him in the summer of 1914 and returned to Zurich with his two children. World War I kicked off with the assassination of Archduke Franz Ferdinand, and the entire continent was set ablaze with war in a complicated web of military alliances.

By the summer of 1915, Einstein had hit the low point of his entire life. Abandoned by his wife and children and malnourished and sickly from lack of food in wartime, Einstein was quickly withering away, his beloved gravity theory withering with him. But all those years of mathematical study and training had not been in vain. In a final grasp of hope, Einstein took one more hard look at his original mathematical approach, the one he had attempted in 1912 – his field equations representing spacetime curvature with Riemann and Ricci tensors.

This time he discovered something remarkable he had missed before. *The multiple solutions he had calculated for his field equations were in fact alternate versions of a single solution - they were equivalent.*

Through pure excitement and sheer force of will, Einstein battled through poor health to use his Ricci and Riemann tensors to calculate the perihelion shift of Mercury. On November 18, 1915, Einstein announced his results – 42.9 arc seconds (Egdall, 222). *This was exactly what astronomers observed via their telescopes.*

But what about length contraction of the distance between two bodies exerting gravity? Einstein realized he didn't need a "global distance" between two objects (like the sun and Earth) because at the most fundamental level, spacetime curvature doesn't occur globally, but locally. (Recall the potato and the fishnet.) The spacetime interval curvature is transmitted from one local frame to another local frame between the sun and Earth, and it is only through the "stitching together" of these countless local frames that a "global" gravitational field is produced.

Finally, he examined his equations again to measure the speed of spacetime curvature travel from place to place – *just as he had predicted, he showed that gravity does not travel instantaneously but moves at the speed of light.*

Ten days later on November 25, 1915, Einstein presented his theory to the Prussian Academy of Sciences, announcing, "Finally the general theory of relativity is closed as a logical structure. The general laws of nature hold true for all systems of coordinates. The theory is of incomparable beauty."

His equations passed all his tests:

They matched Newton's results for slow speeds and weak gravity.

They obeyed the conservation of mass-energy principle.

The equations were covariant, meaning they applied in any and all reference frames.

And for good measure, Einstein had finally explained the mystery of Mercury's strange orbit.

Beside himself with joy, Einstein wrote to his 11-year-old son, saying, "I will try to be with you for a month every year so that you will have a father who is close to you and can love you.. In the last few days I completed one of the finest papers of my life. When you are older I will tell you about it."

In the next chapter, we'll unveil the majesty of his General Relativity theory, the masterpiece of his life.

18

EINSTEIN'S GREAT MASTERPIECE REVEALED!

"Mass-energy grips spacetime, telling it how to curve. Spacetime grips mass-energy, telling it how to move."

— *JOHN ARCHIBALD WHEELER*

Newton conceived of gravity as a mysterious external force acting over distance.

But in Einstein's conception of the universe, space and time are connected in four-dimensional spacetime that is all around us. Massive bodies (as well as energy) cause curvatures in spacetime which we *experience* as gravity. The curvature of spacetime causes things to gravitate toward massive objects, like a marble rolling toward a bowling ball in a bedsheet. The bedsheet nearest the bowling ball is curved most dramatically, and the curvature of the bedsheet becomes less and less noticeable with distance from the bowling ball. *This is the secret of how gravity acts over distances and why length contraction is not an issue. Mass-energy causes local spacetime to change locally, from one small local frame to the next and the next, propagating outward through countless "stitched frames" to produce what we experience as gravity.*

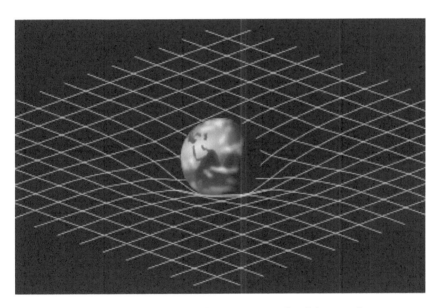

Fig. 16: Earth contains enough mass-energy to bend the spacetime enveloping it, which we experience as gravity. Image Credit: Mysid, licensed under CC By 4.0.

Spacetime curvature *is* gravity. Mass and energy *create* gravity.

In its simplest form, Einstein's General Relativity states:

Spacetime curvature (gravity) = Mass-energy density (source of gravity)

Formally, this **General Relativity Equation** is a set of ten second-order differential equations that, if written out in full detail, would fill an entire encyclopedia with small, tight print. Even today, supercomputers are required to solve the non-linear differential equations of General Relativity. Though it is beyond the scope of this book to fully explain Einstein's General Relativity equations, we will cover them here in their simplest form.

Simply written, it is this:

$$G_{ab} = KT_{ab}$$

Eq. 12: Einstein's General Relativity Equation, where:

G_{ab} = ten-component Curvature Tensor representing space-time curvature

K = constant equal to $\frac{8\pi G}{c^4}$, where G is the gravitational constant

T_{ab} = ten-component Momenergy Tensor representing mass-energy and energy-density

The left side of the equation is the **Curvature Tensor**, which represents spacetime curvature or gravity and how it changes in space over time. It contains the global spacetime interval change within it. The **Momenergy Tensor** on the right side of the equation represents the *sources* of gravity. "Momenergy" is an umbrella term meant to define all possible sources of gravity, which include:

1. Mass density, or the amount of mass contained within a certain volume of space.
2. Energy density, amount of energy contained in a volume of space. (As Einstein taught us, mass and energy are interchangeable and so energy produces gravity too.)
3. Momentum density, or movement of matter and energy through space. Since momentum requires movement and since movement requires energy, momentum is *also* a source of gravity.

Einstein had finally done it. After eight years of struggle, he had finally replaced Newton's gravity theory with his own General Relativity (Siegel, 2021).

But many in the academic elite, unfortunately, were slow to appreciate his genius.

Why should we be so quick to accept Einstein's defiance of Newton? they asked.

Especially after over 200 years of successful observation and proof.

And so it was that Einstein's life's work, his incredible new gravity theory, was not initially met with thunderous applause but with doubt and skepticism (many physicists of the time had not even embraced his Special Relativity theory yet, let alone his expanded General Relativity).

Einstein realized that in order to *prove* his theory beyond reasonable doubt he would need to do more than explain the slight anomaly of Mercury's orbit – he needed to do something bigger, and he needed to do it as quickly as possible.

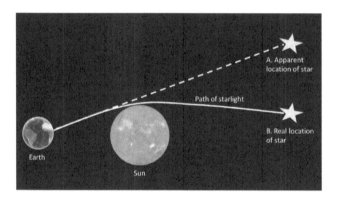

Fig. 17: The Hyades cluster is located behind the Sun with respect to us. As its starlight passes the Sun on its way to Earth, it is warped such that its observed position is shifted from its actual position. Image Credit: Hanoch Gutfreund, licensed under CC By 4.0.

In his third lecture at the Prussian Academy of Sciences, when he was unveiling his masterpiece to the world,

Einstein recalled that in addition to calculating the exact shift of Mercury's perihelion orbit, he had *also* shared that the sun's gravity causes starlight from the Hyades cluster to bend as it moves close to it. In fact, he had even calculated and predicted the exact amount of starlight shift an astronomer would observe as light from stars located behind the sun *bent around the sun on its way to the Earth* – 1.75 arc seconds (Einstein, 92).

But how could anybody actually measure the starlight shift? The sun is too bright in the daytime to observe the stars, and the sun is not visible at all in the nighttime.

What he needed, Einstein realized, was an eclipse.

19

WE NEED PROOF!

> *"It is better to keep your mouth closed and let people think you are a fool than to open it and remove all doubt."*
>
> — MARK TWAIN

A hushed excitement filled the air in London on November 6, 1919 as the Royal Society waited for news from astronomer Arthur Eddington's expedition inside the Burlington House. Earlier that summer he had traveled to Principe in the Gulf of Guinea to observe the solar eclipse of 1919 to prove Einstein's prediction that the sun would shift starlight from the Hyades cluster (located behind the sun) by exactly 1.75 arc seconds – double what Newton would have predicted. Eddington shrewdly sent a back up expedi-

tion to the village of Sobral in the Amazon forest of Brazil to photograph the eclipse as well in case of bad weather.

Einstein himself was sick with anxiety and anticipation as he waited for the news in Berlin. This expedition had been eight years in the making. The first solar eclipse expedition in 1912 had been sabotaged by bad weather, so was the next solar eclipse expedition in 1914, and so was the next solar eclipse expedition in 1916.

"It's just as well," Einstein thought to himself. "In the early days, my equations were incorrect. If those expeditions had been successful, they would have disproved my early predictions and potentially sabotaged my entire theory – and my reputation to boot."

In the Burlington House, an announcer strode to the podium and declared, "I have the results."

The excited chatter in the audience fell to complete and utter silence.

First, we have news from the Principe expedition. They have recorded a displacement of the Hyades cluster of 1.60 +/- 0.31 arc seconds.

Gasps erupted from the listeners.

"And now we have news from the Sobral expedition they have recorded a displacement of 1.98 +/-0.12 arc seconds."

Could it be? Einstein had predicted a shift of 1.75 arc seconds – right inside the confidence interval of both measurements.

J.J. Thomson, the Royal Society president and the man who discovered the electron, famously turned to the portrait of Sir Isaac Newton hanging on the wall and said, "Forgive us, Sir Isaac, your universe has been overturned."

Reporters and writers and physicists scurried everywhere in hurried excitement.

"Someone, please telegraph Einstein!"

Lorentz, the very same physicist who derived the Lorentz transform, telegraphed the news to Einstein, and overnight Einstein became an international celebrity. The leading British newspaper *The Times* ran a headline reading "Revolution in Science – New Theory of the Universe – Newtonian Ideas Overthrown." His face became perhaps the most well known and recognized face in the world – whether the crowds truly understood his theories or not, they adored him.

A couple years later, American astronomer William Wallace Campbell conducted another solar eclipse expedition to the Kimberly region of Australia to observe the solar eclipse of September 21, 1922 (believing that perhaps the previous measurements were faulty and untrustworthy). With even more advanced and sophisticated equipment, he recorded star positions with an average displacement of 1.72 +/- 0.11 arc seconds, closer to Einstein's prediction of 1.75 arc seconds than any other expedition.

There were no longer any doubters.

Einstein spent the next several years traveling abroad before finally emigrating to the U.S. in 1933 to escape the rise of the Nazis and Adolf Hitler. Seven years later in 1940, he would become an American citizen.

In the next chapter, we'll take one last look at the most foundational and important piece of his General Relativity theory and explore why it is so incredibly bizarre. And we'll do it with one last famous experiment – the **spinning bucket**.

20

SO WHAT?

"With fame I become more and more stupid, which of course is a very common phenomenon."

— ALBERT EINSTEIN

It's tempting to look at some of the basic ideas in Einstein's General Relativity and protest, "So what? I could have thought of that. Not special." For example, let's take another look at the most important central piece of his General Relativity:

All laws of physics are the same in all reference frames.

On the surface, this sounds like common sense, but assuredly it is a very strange truth indeed.

To appreciate how insane this idea truly is, let's examine Newton's original spinning bucket experiment. Over 300 years ago, Newton suspended a bucket from a tree with a rope, filled the bucket with water, twisted the rope very tightly, let go, and watched the bucket spin. Slowly and steadily, he watched the water inside the rotating bucket climb up the sides of the bucket, forming a concave depression in the center.

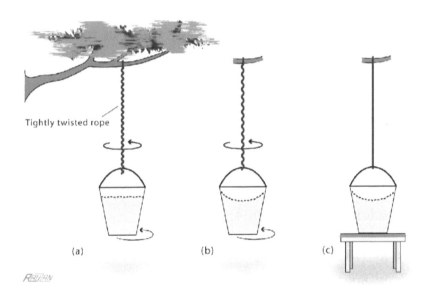

Fig. 18: Newton's spinning bucket experiment. Spinning the bucket causes the water surface to warp into a concave shape. Image Credit: Conrad Ranzan, licensed under CC By 4.0.

Then, Newton asked a seemingly simple question:

What is the bucket spinning with respect to?

According to his Ether theory, the universe is at absolute rest. Therefore, he reasoned, the bucket is spinning with respect to the universe, which is absolutely at rest.

But Einstein tells us there is no such *thing* as "absolute rest." According to his General Relativity theory, all reference frames are valid, and all physics laws are the same in all reference frames. So what does this mean for Newton's spinning bucket?

There are two possible frames of reference to observe the spinning bucket: one is the ordinary frame of reference of sitting outside the bucket and watching it spin. The other *equally valid* reference frame is to perceive the bucket is perfectly still *while the entire universe is spinning around it*. Einstein's incredible theory asserts that both scenarios are equally valid and both scenarios would indeed cause the water inside the bucket to form the exact same concave depression (O'Connor, 2004).

Strange Facts: *Physicists Joseph Lense and Hans Thirring later built on Einstein's General Relativity theory in 1918 to say that just as objects warp space and time, a rotating object can drag space and time with it in a process known as* **frame dragging.** *Later calculations and estimates of the total mass-energy contained within our universe confirmed that according to the equations of General Relativity and frame dragging, a rotating* *universe surrounding a stationary bucket filled with water would indeed "drag" the spacetime contained inside the bucket, producing the exact same concave depression we would see inside a spinning bucket.*

The truth, it would seem, is indeed stranger than fiction. In the next several chapters, we'll discuss several strange and bizarre theories and discoveries that flowed out of Einstein's General Relativity (as well as his single greatest regret) as he raced to complete the task that would haunt him for the rest of his life - uniting Quantum Physics, the physics of small things, with his General Relativity to form a unified Theory of Everything.

21

UNKNOWN GENIUS SOLVES EINSTEIN'S EQUATIONS, DISCOVERS A MIRACLE

"Black holes might be useful for getting rid of garbage or even some of one's friends."

— STEPHEN HAWKING

Karl Schwarzchild sat in his hospital bed and couldn't take his eyes off his physics journal. The year was 1915, and Einstein had just delivered his theory of General Relativity in just four lectures, overturning centuries of Newtonian thought about gravity. Karl devoured a written summary with breathless excitement. It was a welcome distraction.

He had volunteered to serve in the German Army a few short months earlier and was sent to the Eastern Front as a Field Artilleryman. He didn't have to do it, but Karl knew he was Jewish himself and saw the writing on the wall – as antisemitism surged across Germany, he had no choice but to serve in the army to prove he was just as German as every other German.

But in a twist of fate, Karl was besieged with Pemphigas vulgaris, a rare autoimmune disease which caused his immune system to attack his own skin. Blisters spread all over his body, and he was forced to leave the front to receive treatment in a field hospital. He soon turned to physics to distract him from his suffering. Karl happened to be a brilliant physicist and was one of the few people alive who could truly decipher and appreciate the incredible beauty of Einstein's 10 differential equations describing gravity and how mass causes spacetime to curve around it.

"How does gravity work around a star?" Karl wondered.

"Well, a star is a sphere, is it not?"

"Wait a minute," Karl muttered to himself. He put his journal to the side for a few minutes and began furiously jotting down calculations on scratch paper.

"I've got it!"

Miraculously, Karl had finished a solution of Einstein's field equations for how spacetime curves around a perfectly spherical object in space - in other words, he had calculated the gravity of a star. He quickly sent a letter to Einstein in Berlin with his solution included.

But Karl wasn't done. "Stars burn through their fuel until fusion can't happen anymore, and then they collapse from the force of their own gravity. And as stars collapse, they become denser and denser as more and more mass is crushed into a tinier and tinier volume, which creates even stronger gravity. When does it end?"

Eventually, Karl realized, the star would collapse into an infinitely tiny point, a singularity whose gravity would be so great that nothing, not even light itself, could escape. He formulated the mathematical basis of such a black hole and sent his work again to Einstein in Berlin. Einstein received the work and admired the mathematical solution of his field equations, although he did not actually believe nature was capable of producing such a strange object. A few months later in the spring of 1916, Karl Schwarzchild died.

Over fifty years later in 1971, a young astronomer named Paul Murdin was in search of X rays in the skies. X rays, Murdin knew, could only be emitted from matter heated to millions and millions of degrees Fahrenheit, and so he intuited that anything in the cosmos capable of emitting them must be worth studying. NASA had just launched Uhuru, the first X ray satellite, and the results had just come back from its observations. Murdin was beside himself with excitement.

After scanning the results, Murdin saw there was one major X ray source, named Cygnus X-1, inside the Cygnus constellation, and inside this field was a blue supergiant star that was many, many times more massive than our sun.

"The star isn't releasing the X rays," Murdin thought to himself. "Perhaps it's orbiting something that is?"

Murdin turned to his colleague Louise Webster and asked her to measure the velocity and movement of the blue supergiant – sure enough, she observed that it was orbiting something invisible, completing one revolution every 5.6 days. Even more incredibly, the speed of the supergiant's movement was so great that Murdan and Webster calculated that the mass of whatever it was orbiting must have been at least six solar masses. White dwarfs and neutron stars, which had only been recently discovered within the past decade, were not massive enough to explain this kind of behavior. Theoretically, only a black hole could generate such gravity - Murdin and Webster had accidentally stumbled on one.

Today, black holes are common knowledge and we see them everywhere in our pop culture. Just a few years ago, in 2019 the Event Horizon Telescope (EHT) collaboration released the very first image of a black hole in space. The EHT spotted the black hole in the center of galaxy M87 while the telescope was examining its **event horizon**, or the area past which nothing can escape from a black hole. Even with this incredibly powerful telescope, black holes are still nearly impossible to detect because of their distances from Earth and their small size.

Precisely, black holes themselves cannot be observed. Instead, we are only able to observe the effects and shadows and glow *surrounding* the black holes. These effects include the galaxies and stars which orbit at high speeds around seemingly empty and dark areas of space, the superheated disks of matter that glow due to the immense forces, and the circular shadow produced as the black hole bends light around itself.

In 2019, the EHT observed a circular shadow inside galaxy M87, of light bending around a black hole. And in May of 2022, the same EHT finally published an image of Sagittarius A, the supermassive black hole at the center of our own Milky Way. Sagittarius A, located 27,000 lightyears away, is smaller than the size of Mercury's orbit yet contains a mass equivalent to 4 million suns – this discovery directly confirms the prediction stemming from Einstein's General Relativity that superdense objects in space, such as black holes, generate enough gravity to bend spacetime and pull in surrounding light, dust, and matter.

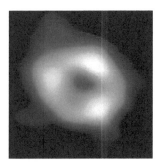

Fig. 19: Black Hole (Sagittarius A) discovered and imaged at the heart of the Milky Way, May 2022. Image Credit: EHT Collaboration, licensed under CC By 4.0.

Strange Facts: *If you fell into a black hole, physicists have long surmised that gravity would stretch you out like spaghetti, though your death would come long before you reached the black hole's center. But a 2012 study published in the journal Nature suggested that quantum effects would cause the event horizon to act much like a wall of fire – you wouldn't be stretched to death. You would burn to death. Any surrounding light would be warped by the incredible gravity of the black hole, and at the event horizon the light wave would be stretched perfectly flat, making its frequency zero. And as we revealed in earlier chapters, frequency is a marker of time. Thus, time itself is frozen at the event horizon.*

So how does such a monster form in the real world? As Karl Schwarzchild guessed in 1916, a stellar black hole forms when a star that is about 20 times more massive than the sun runs out of nuclear fuel. At this point, the outward force of its dying nuclear engine is no longer strong enough to resist the unrelenting crush of its own gravity. Consequently, the star implodes, its core collapses, and the outer layers rebound outward in a massive supernova explosion.

If the core's mass is less than twice the mass of the sun, the imploding core becomes a neutron star. But if the core's mass is greater than two solar masses, the imploding star's gravity is great enough to overcome all subatomic forces holding its atoms together and it undergoes an endless collapse to an infinitesimally small point, a singularity. The distance between a black hole's singularity and the event horizon, the point past which nothing can escape, is called the Schwarzschild radius.

Observation of a jet emanating from a black hole. Image Credit: NASA, JPL-Caltech/IPAC, licensed under CC By 4.0.

Billions of galaxies in our universe are actually rotating around supermassive black holes formed near the birth of our universe. These spinning black holes are so massive they are able to cause spacetime around them to spin as well through frame dragging, causing in-falling material to rotate millions of miles an hour around them. This, combined with the incredible gravitational force of the black hole, causes the debris surrounding the black hole to rise to incredibly high temperatures, and great twisting magnetic fields cause this hot material to shoot outward in jets at nearly the speed of light (Tillman, 2023).

Readers may now appreciate the miraculous importance of black holes – nothing else in the cosmos unites the physics of Big Things with Small Things, of General Relativity with Quantum Theory, like black holes. Here is a supermassive object whose gravity is so strong that light can't escape its grasp and entire star systems can't escape its orbit, but also an object which contains a quantum singularity at its center where gravity and density are so high that spacetime itself breaks down. To understand a black hole completely is to understand quantum gravity, and thus uncover a Theory of Everything.

ON WORMHOLES

Before we continue the story of Einstein's adventures, let's briefly examine another one of Einstein's predictions that has fascinated science fiction fans for decades – wormholes. If spacetime is like a fabric that can bend, what would happen if the fabric folded over itself? Could it be possible to instantly travel vast distances in the universe, simply by moving through these folds in spacetime? Schwarzchild had solved Einstein's equations to predict the existence of black holes in space – could not such an object also fold spacetime?

In 1935, Albert Einstein and Nathan Rosen predicted the existence of "Einstein-Rosen bridges," popularly known as wormholes, and theorized that objects in space with immense gravity, such as black holes, could indeed create temporary wormholes in spacetime. The simplest way to explain a wormhole tunnel is to imagine a piece of paper as the fabric of spacetime. Fold the paper in half and stab a pencil through it so that the pencil pierces both sheets. You have a black hole on one side, through which nothing can escape, and a "white hole" on the other side, through which nothing can enter, with a wormhole joining them. Unfortunately for science fiction fans, Einstein and Rosen predicted two major problems that would render wormholes useless as a means of future space travel. First of all, they would be incredibly tiny and invisible to the naked eye. Secondly, they would also be incredibly unstable, likely collapsing merely moments after formation, because any matter passing through a wormhole would have a gravitational force of attraction that would inevitably pull the walls of the wormhole shut (Redd, 2017).

In the next chapter, we'll reveal Einstein's greatest blunder, told through the incredible true story of a teacher, a janitor, a Catholic priest, a Russian mathematician, and the great patent clerk himself, Albert Einstein.

22

THE TEACHER, THE JANITOR, THE PRIEST, THE RUSSIAN, AND THE PATENT CLERK

"Equipped with his senses, man explores the universe around him and calls the adventure Science."

— EDWIN HUBBLE

By 1917, Einstein had already developed his field equations for General Relativity, and he had already used them to accurately explain the mystery of Mercury's orbit. But something else troubled him greatly. Believing (along with all the other physicists of the time) that the entire universe was contained in just our Milky Way galaxy, Einstein endeavored to calculate the spacetime curvature of the entire known universe, based on estimates of all the mass-energy observed in the known universe.

His results bothered him deeply. What his equations revealed was that a universe containing mass-energy cannot stay the same size – it must either grow or contract. This collided with the prevailing scientific wisdom of the day, of course, which dictated that the universe is static. The universe, according to physicists of the time, had always been the same exact size and would always *be* the exact same size.

"What is wrong with my equations?" Einstein thought to himself.

To fix the problem, Einstein pulled a clever mathematical trick and added a **cosmological constant** to his General Relativity equations, one which would apply outward force to counteract the gravity of the universe so that the net result would be that the universe would remain perfectly still, in accordance with the wisdom of the times. Satisfied, he put his notes and calculations to the side and shifted his thoughts to other things.

Five years later in 1922, as the Soviet Union was established around him, Russian mathematician Alexander Friedmann published several solutions of Einstein's General Relativity equations (without the cosmological constant) and showed that the universe "can expand, contract, collapse, and might even be born in a singularity." Einstein doubted Friedmann's work, calling it "suspicious." Three years later, Friedmann died from typhoid fever at 37.

Two years later in 1927, operating without *any* knowledge of Friedmann's published work, a Roman Catholic priest from Belgium named Abbe Georges Lemaître independently solved Einstein's General Relativity equations and also realized that the universe is dynamic and can change size. But he took it a step further. Borrowing from Edwin Hubble's discovery in 1924 that there are billions of other galaxies outside our own Milky Way galaxy, Lemaître proposed that the light from distant galaxies undergoes decreasing frequency (or *redshift*) as it travels to Earth, and there is a proportional relationship between a galaxy's distance from Earth and its redshift.

In other words, the further away a galaxy is, the lower its light frequency will be as detected from Earth. Lemaître's next brilliant insight was to suggest that *a distant galaxy's light is stretched in frequency because space itself is expanding. The universe is expanding.*

The same year, Lemaître attended the prestigious Solvay Conference in Brussels, where he met Einstein, his idol, in person. Barely able to contain his excitement, Lemaître showed his calculations to Einstein, explaining that not only is the universe dynamic, but that it is *expanding*.

"Your calculations are correct," Einstein replied. "But your grasp of physics is abominable."

Two years later in 1929, astronomer Edwin Hubble, working alongside his assistant, astronomer Milton Humason (a former janitor), measured the distances to 24 galaxies with the best methods available at the time while Humason measured their redshifts, and what he found mirrored exactly the results Lemaître had produced years before – the galaxies he observed were moving further and further away from each other.

In his famous landmark 1929 paper, Hubble gave us Hubble's Law, which gives us the precise *velocity* of a galaxy that is moving away from us. As you can see, *not only are galaxies moving away from us, but the galaxies that are further away from us are moving at a much greater velocity than the ones that are closest to us.*

$$v = H_0 d$$

Eq. 13: Hubble's Law

In this equation,

v = receding velocity of a star or galaxy

H_0 = Hubble's constant, 70 km/s/Mpc (where 1 Mpc = 10^6 parsec = 3.26×10^6 lightyears) *d* = estimated distance from Earth

By rewriting his Law, Hubble was able to extrapolate different Hubble times, or ages, of the universe for different Hubble constants. Given the present Hubble constant shown in the equation above, Hubble calculated that the universe began its expansion roughly 13.8 billion years ago. In one leap, Hubble overturned the Static Model of the universe, showed that the universe is expanding, and he calculated its age (Bahcall, 2015).

Strange Facts: *While we know the universe is 13.8 billion years old, archaeological evidence shows that Life on this planet emerged only 3.7 billion years ago (and on five separate occasions it has nearly been wiped out). Following the most recent mass extinction event roughly 65 million years ago, a deadly meteor strike that wiped out the dinosaurs, our own species, homo sapiens, gained consciousness roughly 70,000 years ago and has since become the most dominant and dangerous life form our planet has ever known. Given we will most certainly face another mass extinction event in the future, physicists assert we must evolve from a Type 0 civilization to a Type I civilization, one which is capable of altering and controlling weather, deflecting space debris, harnessing energy from the sun, and (if necessary) leaving our planet entirely. Indeed, the future survival of our species depends on how quickly we are able to make this leap. Morbidly enough, the need to advance fast enough to overcome the next mass extinction event is one that* *applies equally to all intelligent Life in the universe, and while the Law of Big Numbers says intelligent Life is statistically certain to exist elsewhere in the cosmos, this morbid hurdle limits its growth and darkly answers Fermi's famous question: Where are all the aliens? As evidence of UFO sightings continues to emerge from credible military and government sources, however, public opinion on this matter is beginning to shift as we march through the 21st century. The Aliens, it would seem, are already here.*

Einstein, confronted with the incredible work of both Hubble and Lemaître, was forced to concede that he was wrong – we do, in fact, live in an expanding universe. He had stumbled on this fact a decade earlier in 1917, but he had allowed his own prejudices and biases to dissuade him from accepting it. Thus, he lamented that his cosmological constant was indeed the "greatest blunder of [his] life." In 1931, he personally visited Edwin Hubble to congratulate him and to thank him for delivering him from folly.

Careful readers may be asking a very good question at this point – if a galaxy's speed increases the further away from us it is, wouldn't this speed eventually surpass the speed of light at a great enough distance? Didn't Einstein show us that nothing can ever travel faster than the speed of light? Hubble and Einstein manage to wriggle out of this riddle, saying that nothing can travel through space faster than light, but space itself isn't bound to such rules. The universe is indeed expanding faster than the speed of light in all directions.

You may recall that although Hubble received widespread fame and recognition for his discovery that the universe is expanding, it was Father Lemaître, not Hubble, who originally published the idea, two years earlier than Hubble. But the priest was far from finished. In the next chapter, we'll reveal how Lemaître used the expansion of the universe to explain something else – its beginnings. The implications of his idea would send shockwaves through the world, impacting scientists, philosophers, and even the Pope himself.

23

IN THE BEGINNING

"Man was made at the end of the week's work when God was tired."

— *MARK TWAIN*

"Good heavens," Father Lemaître thought to himself.

After sharing his expanding universe ideas with Einstein in 1927, the great patent clerk had shared Friedmann's dynamic universe papers with the priest. Four years later, Father Lamaitre found his mind wandering again and again back to the vision of an expanding universe.

"If the universe is indeed expanding, it must have been smaller in the past."

The priest excitedly rushed to his desk to write down his thoughts.

"What if the universe began as a single quantum? We could conceive the beginning of the universe in the form of a unique atom, the atomic weight of which is the total mass of the universe."

Lemaître spent the next several years spreading his idea in lectures, receiving both praise and criticism wherever he went. Einstein called his theory the "most beautiful and satisfactory explanation of creation to which [he had] ever listened." Pope Pius XII latched onto the theory as proof of the biblical Genesis story. Father Lemaître himself adamantly rejected the notion that his idea confirmed or disproved religious beliefs. It was only a scientific matter – nothing more.

Several years later in 1948, physicists George Gamow and Ralph Alpher extended Lamaitre's ideas to say that the universe came into being through a massive explosion of space roughly 13.8 billion years ago, known today as the **Big Bang.**

Alpher took the theory even further and made a *measurable prediction* (a bold move for any theory). According to Alpher, the early universe was too hot and dense for matter atoms to form. But about 380,000 years after the Big Bang, the universe cooled enough to form neutrally charged atoms out of the soup of free ranging electrons, photons, neutrons, and protons. Suddenly, these primordial photons, which had previously been absorbed by other free electrons and protons, were now free to roam the universe as light, thus filling the universe with high-energy light.

But over billions of years as the universe continued to cool and expand, this high-energy primordial light began to lose energy as its frequency was stretched longer and longer over time. Thus, predicted Alpher, the light from the early universe exists today as low-energy microwaves that can be observed all over the universe, and the temperature of this cosmic microwave radiation, according to Alpher, is about 5 degrees Kelvin.

Nearly 20 years later in 1965, physicists Arno Penzias and Robert Wilson from Bell Labs were studying radio emissions from the Milky Way with an ultrasensitive microwave receiver when they detected an incessant *hiss* that just wouldn't go away which could be heard all over the skies.

What could it be?

They contacted the lab at nearby Princeton University and shared the news with physicist Robert Dicke, who, serendipitously, was leading a team of physicists in search of Alpher's predicted **Cosmic Microwave Background (CMB).**

"Well, boys," he told his team. "We've been scooped."

Penzias and Wilson measured the temperature of the microwave radiation at 3 degrees Kelvin, just slightly less than Alpher's predicted temperature of 5 degrees Kelvin.

Today, modern physicists fully accept the Big Bang theory based on expansive evidence: the CMB and the fluctuations in the CMB, the age of stars, the presence and amount of hydrogen, helium, and other light elements in the universe, the large-scale structure of the universe, and several other measurements which point to its truth (Wright, 2013).

Unfortunately, we don't know *how* the universe banged or what happened at time zero or just before the Big Bang, and it is certainly outside the realm of science to speculate on *why* the Big Bang happened. Is our universe the only one? Are we the creations of a loving Creator or are we highly evolved, accidental creatures in a vast and uncaring universe? Or stranger still, players inside an unimaginably complex simulation? Where is God? And if we were ever lucky (or unlucky) enough to encounter a sufficiently advanced Life form would we even be able to tell the difference?

24

THE DARK SIDE OF THE FORCE

"If you only knew the power of the Dark Side."

— DARTH VADER, STAR WARS EPISODE V: THE EMPIRE STRIKES BACK

Not only do we know the universe continues to expand today – it expands at an ever faster and faster rate in which billions and billions of cubic light years of space are added each and every second. But, scientists wondered, shouldn't the gravity contained within the universe cause it to contract over time? Why does the universe continue to expand? A star, for example, will collapse from the force of its own gravity. Why wouldn't the universe behave the same way?

Clearly, there was an unknown force acting against gravity, a mysterious anti-gravity or **dark energy** (Tillman, 2013). Given the universe's expansion and the force of gravity within the universe, physicists calculated that dark energy must account for about 68% of the cosmos. Truthfully, scientists have never actually *observed* dark energy or have a clue what it's made of. Nonetheless, scientists have not let their ignorance prevent them from proposing a variety of theories on what this intangible collection of dark energy is, including:

1. A fifth fundamental force in addition to gravity, electromagnetism, and the strong and weak nuclear forces. Some scientists have termed this force or field "quintessence" after Greek philosophers.
2. A fundamental element of space itself
3. A central piece of Einstein's theory of gravity Until scientists can get closer to validating one of these hypotheses, all we can say for sure about dark energy is that the universe is expanding at a faster and faster rate, and we believe dark energy is responsible for this expansion (Wells, 2021).

As early as 1933, Swiss-American astronomer Fritz Zwicky spotted distant galaxies and noticed they were spinning around one another far faster than expected, given their observable masses as seen via telescopes. His reasonable conclusion was that the galaxies possessed mysterious, unobserved mass he called dark matter, which was causing them to spin at such speeds. "If this is confirmed," he wrote, "we will find that dark matter is substantially more abundant than illuminating matter" (Clavin, 2020).

However, many in the profession remained unconvinced of Zwicky's findings until the 1970s, when astronomers Kent Ford and Vera Rubin conducted thorough investigations of stars in the neighboring Andromeda galaxy's outer regions. These stars were circling the galactic core far too swiftly, almost as if some invisible material was tugging on them and pushing them along – an observation scientists soon detected in galaxies all around the cosmos (Mann, 2020). Researchers had no clue what this unseen material was formed of.

Unfortunately, dark matter is undetectable to the naked eye because it emits no light or energy. Some astronomers hypothesized that dark matter was made up of tiny black holes or other compact objects that emitted too little light to be observed through telescopes. According to NASA, the results became increasingly odd in the early 2000s, when a satellite telescope called the Wilkinson Microwave Anisotropy Probe (WMAP) set about observing the microwave radiation left over from the Big Bang. By precisely measuring the microwave fluctuations, the WMAP team was also able to calculate with striking precision the composition of the universe, and they determined that while ordinary matter comprises about 5% of the universe, dark matter comprises 24% of the universe and dark energy comprises 71.4% of the known universe (Briggs, 2020).

WHAT IS DARK MATTER MADE OF?

Visible matter, also known as baryonic matter, comprises baryons, which are collective names for subatomic particles such as protons, neutrons, and electrons. Scientists can only anticipate what dark matter is made of. It could be made up of baryons, but it could also be non-baryonic, something different entirely. Most scientists believe dark matter is made up of non-baryonic stuff.

SO WHAT'S THE DIFFERENCE?

Dark energy and dark matter have little in common other than their common adjective, "dark," which doesn't really describe them properly and is only meant to show us that scientists don't fully understand them yet. That said, here are the two main differences between them:

1. Dark matter possesses an attractional force, like gravity, and does not reflect, absorb, or emit light.

2. Meanwhile, dark energy is a repellent force, a kind of anti-gravity, that propels the universe's ever-accelerating expansion (Clavin, 2020).

The significance of dark matter and dark energy is staggering – we live on a tiny spinning rock that is revolving around a very tiny star inside a very small galaxy in a giant universe that is expanding faster than the speed of light. And the trillions of ordinary particles that make up the planets and stars and galaxies inside this universe account for merely 5% of its total mass – the rest of it is hidden from us in the form of dark energy and dark mass.

In the next chapter, we'll revisit one last time the quest that haunted Einstein for the remainder of his life following his revolutionary work on gravity and spacetime – his search for an overarching **Theory of Everything**. And finally, we'll reveal why Physics has hit a frustrating period of stagnation in the late 20th and early 21st centuries (and offer some hopeful thoughts on how we might overcome it).

25

THE SEARCH FOR A THEORY OF EVERYTHING AND FINAL THOUGHTS

"I want to know God's thoughts – the rest are mere details."

— ALBERT EINSTEIN

Almost immediately after Einstein revealed his masterpiece, his new theory of gravity, to the world over a century ago in November 1915, he set his sights on a problem which would plague him for the next several decades till his death, the same grand problem facing modern physicists today – how to unite the separate branches of electromagnetism, Special and General Relativity, and Quantum Physics into a single unified Theory of Everything.

After Einstein himself quite accidentally co-founded Quantum Physics with his discovery of the photon, the force-carrier particle of light, the great physicists of the day including Max Planck, Paul Dirac, and others carried the torch of the quantum revolution forward, ultimately creating one of the most monumental successes in the history of science – the **Standard Model of Particle Physics**.

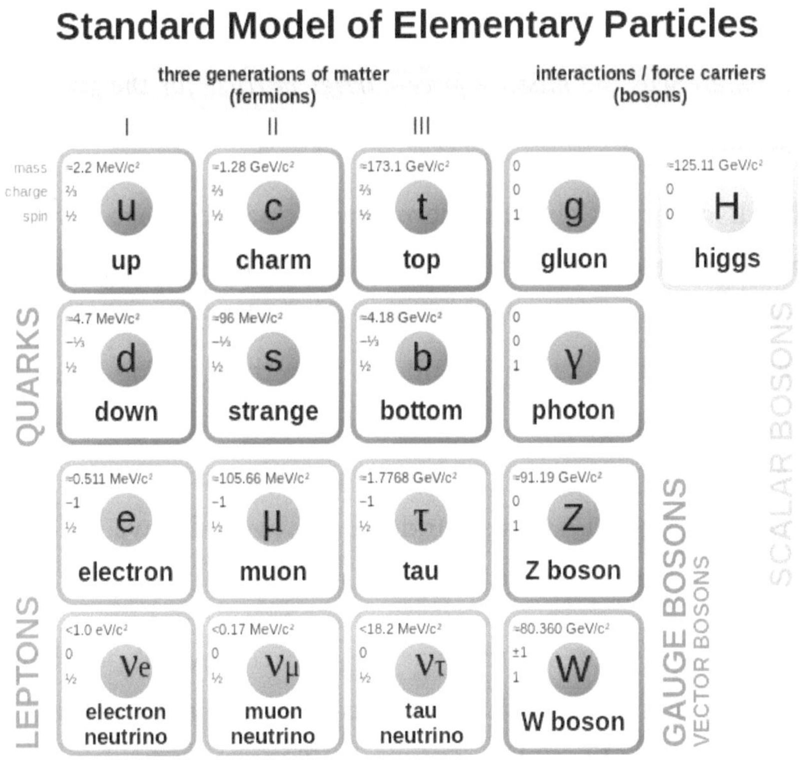
Standard Model of Elementary Particles
three generations of matter (fermions)
interactions / force carriers (bosons)
I
II
III
mass
charge
spin
≈2.2 MeV/c^2 ⅔ ½ u up
≈1.28 GeV/c^2 ⅔ ½ c charm
≈173.1 GeV/c^2 ⅔ ½ t top
0 0 1 g gluon
≈125.11 GeV/c^2 0 0 H higgs
≈4.7 MeV/c^2 −⅓ ½ d down
≈96 MeV/c^2 −⅓ ½ s strange
≈4.18 GeV/c^2 −⅓ ½ b bottom
0 0 1 γ photon
≈0.511 MeV/c^2 −1 ½ e electron
≈105.66 MeV/c^2 −1 ½ μ muon
≈1.7768 GeV/c^2 −1 ½ τ tau
≈91.19 GeV/c^2 0 1 Z Z boson
<1.0 eV/c^2 0 ½ ν_e electron neutrino
<0.17 MeV/c^2 0 ½ ν_μ muon neutrino
<18.2 MeV/c^2 0 ½ ν_τ tau neutrino
≈80.360 GeV/c^2 ±1 1 W W boson
QUARKS
LEPTONS
GAUGE BOSONS
VECTOR BOSONS
SCALAR BOSONS

Fig. 21: Standard Model of Particle Physics. Image Credit: Cush, licensed under CC By 4.0.

This incredible model deconstructed all of reality down into just three classes – mass particles (particles like quarks and protons and neutrons), their antimatter opposites, and force-carrier particles (particles carrying force like photons). By the end of the 20th century, we had successfully discovered all of the mass particles and all of the force-carrier particles for *three* of the four known forces in the universe: force-carrier particles for the electromagnetic force, the strong nuclear force, and the weak nuclear force.

But where was the missing force-carrier particle for the gravitational force? Where was the missing "graviton?"

The discovery of the quantum gravity particle fast became the greatest drama in physics following Einstein's publication of General Relativity – if we could find the missing gravity particle, we would indeed have the missing piece to unite Classical Physics, Special Relativity, and General Relativity (the physics of Big Things) with Quantum Physics (the physics of Small Things). Indeed, we would have a functioning Theory of Everything.

Unfortunately, the search was beset with several problems from the start, including the incredible weakness of gravity (a paperclip holds tightly to a magnet even though it is also being pulled by the gravitational force of an entire planet), as well as the troublesome prediction that a collision of gravitons would release infinite energy, either a sure sign of error somewhere in the mathematics (or alternately, a sign of impending doom).

In the 1960s, several prominent physicists, most notably Edward Witten, proposed **String Theory**, which conveniently and simply solved the gravity particle problem by suggesting that everything in our cosmos is not composed of particles but is instead composed of infinitely tiny one-dimensional *strings*. Unfortunately, this exciting theory, which has monopolized the brightest minds in physics as well as the lion share of funding and research in physics for over fifty years, has yielded little in the way of verifiable, provable results and evidence. Indeed, the theory that was seemingly so simple at first has continued to produce ever-greater theories to justify its own existence, such as the addition of *six* extra "hyperspace" dimensions in addition to the four dimensions of Einstein's universe (Stein, 2023).

100 years after Einstein began a serious lifelong search for the Theory of Everything and a search to solve the quantum gravity problem, we are still grasping and searching for answers.

But, perhaps, this climate is not so very different from the one young Albert Einstein inhabited at the dawn of the 20th century, when the world's best and brightest physicists were stuck in an intellectual rut, endlessly arguing over the finer points of Newton's beloved Ether theory instead of pausing to ask themselves if Ether was even real in the first place.

And, perhaps, somewhere out there in the world there's a young man or lady who seeks to drag us out of the mud of our own ignorance.

Someone who is unburdened by the shackles of conventional science or career or reputation or status quo.

Someone with a talent for day dreams and mischief.

Someone who can show us the way.

ACKNOWLEDGEMENTS AND REVIEWS

My team, Publishing Services, and I have poured our heart and soul into these pages in order to educate, entertain, and inspire our Readers – if you gained Value from reading this book and would like us to publish future books, please let us know by leaving a 5 star review. We appreciate it! Scan the QR code below to leave your review.

AUTHOR'S OTHER WORKS

Quantum Physics for Beginners, Into the Light: The 4 Bizarre Discoveries You Must Know To Master Quantum Physics Fast, Revealed Step-By-Step (In Plain English)

GLOSSARY

Acceleration: change in direction or the rate at which an object changes its velocity

Annihilation: reaction in which a particle and its antiparticle collide and disappear, releasing energy. The most common annihilation on Earth occurs between an electron and its antiparticle, a positron.

Big Bang: the rapid expansion of matter from a state of extremely high density and temperature that according to current cosmological theories marked the origin of the universe roughly 13.8 billion years ago.

Black hole: region of spacetime where gravity is so strong that nothing, including light or other electromagnetic waves, has enough energy to escape it. The theory of General Relativity predicts that a sufficiently compact mass can deform spacetime to form a black hole

Conservation of mass: total mass in an isolated closed system doesn't change.

Conservation of energy: total energy in an isolated closed system doesn't change.

Conservation of mass-energy: total mass-energy in an isolated closed system doesn't change.

Cosmic Microwave Background: microwave radiation that fills all space in the observable universe. It is a remnant that provides an important source of data on the primordial universe.

Cosmological constant: constant coefficient of a term that Albert Einstein temporarily added to his field equations of General Relativity in order to keep the universe perfectly static. He later removed it.

Curvature tensor: represents spacetime curvature and how it changes from place to place over time

Dictum on Uniform Motion: Galileo's proposition that the laws of physics for reference frames moving at constant or uniform speed relative to each other are the exact same as the laws of physics for reference frames at rest relative to each other.

GLOSSARY | 169

Doppler effect: increase or decrease in the frequency of sound, light, or other waves as the source and observer move toward or away from each other. The effect causes the sudden change in pitch noticeable in a passing siren, as well as the redshift seen by astronomers.

Double flashlight thought experiment: thought experiment developed by Albert Einstein in 1905 in which a flashlight emits light from both ends. Einstein used this thought experiment to derive his famous mass-energy equation, ..

Ehrenfest paradox: paradox formulated by physicist Paul Ehrenfest in 1909 highlighting the fact that according to Einstein's Special Relativity, a spinning disc will have a smaller circumference than its rest circumference because length contraction occurs in the direction of the disc's movement.

Electromagnetic waves: synchronized oscillations of electric and magnetic fields propagating through space at the speed of light c. Includes radio waves, microwaves, infrared, visible light, ultraviolet, x-rays, and gamma rays.

Elevator light thought experiment: ingenious thought experiment that shows that light must bend in a gravitational field. An elevator in free fall contains a laser on one side that fires a photon to a detector on the other side and two observers watch the photon. The person inside the elevator watches the photon travel in a perfectly straight line from the laser to the detector. But the person standing on the ground observes that the photon follows a slightly *curved* path from the laser to the detector as the elevator moves in free fall.

Equivalence principle: a gravitational force causes the exact same physical effects as acceleration .

Energy: the ability to do work. Exists in potential, kinetic, thermal, electrical, chemical, nuclear, or other forms. Energy can neither be created nor destroyed but only changed from one form to another.

Ether: theoretical universal substance believed to act as the medium for light and sound and believed to constitute absolute rest. Adopted by Sir Isaac and other scientists at the beginning of the 17th century, this theory was weakened by the Michelson-Morley experiment of 1887, which showed that light moves at constant speed. Einstein's Special Relativity theory in 1905, which showed that absolute rest doesn't exist and established the speed of light as a universal constant, rejected the Ether theory entirely.

Event horizon: boundary around a black hole beyond which no light or other radiation can escape.

Frame dragging: a phenomenon based on General Relativity in which rotating mass "drags along" spacetime in its vicinity, also known as the Lense-Thirring effect.

Frequency: the number of waves that pass a fixed point in unit time, measured in Hertz. As the energy of an electromagnetic wave increases, so does its frequency.

Galilean transform: used to convert the coordinates of two reference frames which vary only by constant relative motion.

Gaussian coordinates: coordinate system for Euclidean space in which the coordinate lines may be curved.

General Relativity equation: spacetime curvature is equal to mass-energy density

Kinetic energy: form of energy an object possesses due to its motion, measured in Joules. It is equal to half the mass of an object multiplied by the square of its velocity.

Length contraction: phenomenon in which an outside observer measures the length of a moving object to be shorter than the object's proper length, the length measurement in its rest frame.

Light clock thought experiment: Two Light Clocks contain mirrors with a single photon bouncing up and down between them at the speed of light. One Light Clock is moving with respect to the other stationary Light Clock. Observers in both reference frames will observe their photon is bouncing up and down for their individual Light Clocks. But an observer at rest relative to a moving Light Clock will observe that the photon of the moving Light Clock must travel a longer distance as it bounces between mirrors which are uniformly moving. Since the speed of light is the same for all observers, constant speed of the photon coupled with the longer distance it must travel requires the time interval to increase for the moving Light Clock. Therefore, time "slows down" in the moving reference frame relative to the stationary reference frame.

Light Postulate: The speed of light is absolutely the same in all reference frames, independent of the speed of the light source and independent of the speed of the observer.

Manhattan Project: U.S. government research project (1942-1945) to build the first nuclear weapons in response to the threat of the Nazis and Axis Powers in WWII. On July 16, 1945, scientists used the nuclear fission of uranium-235 to detonate the first atom bomb in Alamogordo, New Mexico.

Momenergy tensor: , represents the sources of gravity, including mass/energy/momentum density

Momentum: measure of an object's motion, equal to the product of an object's mass and its velocity.

Newton's Laws of Motion: Newton's first law states that if an object is at rest or moving at constant speed in a straight line, it will continue to stay at rest or move in a straight line at constant speed until it is acted on by an outside force.

Newton's second law that an object's force is equal to the product of its mass and acceleration. Newton's third law states that when two objects interact, they apply forces to each other that are equal in magnitude but opposite in direction.

Nuclear fission: reaction in which the nucleus of an atom splits into two or more smaller nuclei. The fission process often produces gamma photons, and releases a very large amount of energy even by the energetic standards of radioactive decay. The atom bomb uses the nuclear fission of uranium-235 to release massive energy.

Particle-wave duality: central concept of quantum physics asserting that all matter has wave-like properties and all particles have wavelike properties. Discovered through the work of Max Planck, Albert Einstein, Louis de Broglie, Neils Bohr, Schrödinger, and others.

Photon: force carrier particle of the electromagnetic force. Photons have no mass, no charge, and travel at the speed of light.

Principle of covariance: the laws of physics remain the same in all reference frames.

Proper time: the time measured by a clock in an observer's reference frame, or the time measured by a clock with the same motion as observer.

Quantum Field Theory: combining the elements of quantum mechanics with those of relativity to explain the behavior of subatomic particles and their interactions via a variety of force fields.

Quantum physics: study of matter and energy at the atomic and subatomic level.

Radioactivity: phenomenon of the spontaneous disintegration of unstable atomic nuclei to form more energetically stable atomic nuclei, releasing electromagnetic radiation in the process.

Reference frame: set of coordinates that can be used to determine positions and velocities of objects in that frame.

Relativity Postulate (Special Relativity version): all laws of physics are the same for all uniformly moving reference frames.

Simultaneity: concept that distant simultaneity – whether two spatially separated events occur at the same time – is not absolute, but depends on the observer's reference frame

Space elevator thought experiment: a man in an elevator under a gravitational force will experience the same physical effects as a man accelerating upward in an elevator in zero gravity outer space with acceleration .

GLOSSARY | 175

Special Relativity: scientific theory of the relationship between space and time. In Albert Einstein's original treatment, the theory is based on two postulates: 1) the laws of physics are invariant (identical) in all inertial frames of reference (that is, frames of reference with no acceleration); 2) the speed of light in vacuum is the same for all observers, regardless of the motion of light source or observer.

Spinning bucket thought experiment: there are two possible frames of reference to observe Newton's spinning bucket. One is the ordinary frame of reference of sitting outside the bucket and watching it spin. The other *equally valid* reference frame is to perceive the bucket is perfectly still *while the entire universe is spinning around it.* Einstein's incredible theory asserts that both scenarios are equally valid and both scenarios would indeed cause the water inside the bucket to form the exact same concave depression.

Superposition: a particle can possess many position, energy, and momentum states at once. However, a measurement always finds it in one state, but before and after the measurement, it interacts in ways that can only be explained by having a superposition of different states.

String Theory: theoretical framework in which the point-like particles of particle physics are replaced by one-dimensional objects called strings. String theory describes how these strings propagate through space and interact with each other.

Theory of Everything: singular, all-encompassing, coherent theoretical framework of physics that fully explains and links together all aspects of the universe.

Thought experiment: imaginative, hypothetical situation in which a hypothesis, theory, or principle is laid out for the purpose of thinking through its consequences

Time dilation: the "slowing down" of a clock as determined by an observer who is in relative motion with respect to that clock.

Train paradox: if lightning strikes both ends of a moving train, an outside observer located at the train's midpoint will perceive the lightning bolts as simultaneous as light from the lightning strikes reaches him at the same time. But a separate observer inside the moving train will move toward one of the lightning bolts and away from the other and so the light from one of the lightning strikes will reach him before the other. Thus, he will perceive that one of the lightning strikes occurred "before" the other. Simultaneity is relative.

Twin paradox: thought experiment devised to disprove Einstein's concept of time dilation. The thought experiment asserts that time dilation works in both directions, and so if a stationary twin remained on Earth and a moving twin left

Earth moving at near light speed at constant speed and returned to Earth, each twin would observe time "slow down" for the other, and so it would be impossible to determine which twin would be "younger" upon the moving twin's return. Einstein brilliantly replied that since the moving twin would have to reverse his direction to return to Earth, the acceleration would cancel out his time dilation so that time dilation would only occur for the stationary twin on Earth: time would pass more slowly for the moving twin only and therefore he would be the "younger" twin upon his return to Earth.

Worldline: the path that an object traces in 4-dimensional spacetime.

Wormhole (Einstein-Rosen Bridge): speculative structure linking disparate points in spacetime, and is based on a special solution of the Einstein field equations. A wormhole can be visualized as a tunnel with two ends at separate points in spacetime.

QUICK REFERENCE GUIDE

Special Relativity:

1. Galileo's Dictum: All laws of physics (classical physics) are the same for all reference frames moving in constant motion (constant speed) relative to each other.

2. Relativity Postulate: All laws of physics (including classical physics and electromagnetism and as yet undiscovered physics) are the same for all reference frames moving in constant motion (constant speed) relative to each other.

3. Light Postulate: The speed of light is absolutely the same in all uniformly moving reference frames, inde-

pendent of the speed of the light source and independent of the speed of the observer.

4. Time Dilation: When an object moves, time slows down for the object relative to stationary observers.

5. Simultaneity is relative.

6. Length Contraction: When an object moves, its length contracts in the direction of its movement relative to stationary observers.

7. Mass-Energy Equivalence: The mass of a body is the measure of its energy content.

8. Spacetime Interval: The spacetime interval is absolute and invariant (doesn't change) with relative uniform motion.

General Relativity:

1. Equivalence Principle: Gravity and acceleration have the same physical effects.

a. Acceleration and gravity cause light to bend.

b. Acceleration and gravity affect the frequency of light.

c. Acceleration and gravity affect the energy of light.

d. Acceleration and gravity cause time dilation.

e. Acceleration and gravity cause length contraction.

2. The spacetime interval changes in a gravitational field.

3. Relativity Postulate (expanded): All laws of physics are the same for all uniformly moving reference frames, accelerating reference frames, and gravitational reference frames.

4. Spacetime curvature *is* gravity. Mass and energy *create* gravity.

EQUATIONS

Eq. 1: Newton's Speed Combining Formula

For speeds v_1 and v_2, the combined speed V is:

$$V = v_1 + v_2$$

Eq. 2: Einstein's Speed Combining Formula

For speeds v_1 and v_2, measured as a percentage of the speed of light, the combined speed V is:

$$V = \frac{v_1 + v_2}{1 + v_1 v_2}$$

Eq. 3: Einstein's Time Dilation Equation

$$\Delta t' = \frac{\Delta t}{\sqrt{1-v^2}}$$

$\Delta t'$ = dilated time observed by someone in the other reference frame

Δt = time observed in one's own reference frame (rest time)

v = speed of the moving reference frame as a percentage of the speed of light

$\sqrt{1-v^2}$ = Lorentz Factor

Eq. 4: Einstein's Length Contraction Equation

$$\Delta x = \Delta x'\sqrt{1-v^2}$$

$\Delta x'$ = length of an object at rest relative to an observer

Δx = contracted length of an object in motion relative to an observer

v = speed of an object as a percentage of the speed of light

$\sqrt{1-v^2}$ = Lorentz Factor

Eq. 5: Newton's Momentum Equation

$$P = mv$$

P = momentum

m = mass

v = velocity

Eq. 6: Einstein's Momentum Equation

$$P = \frac{mv}{\sqrt{1 - v^2}}$$

P = momentum

m = mass

v = velocity as a percentage of light speed

$\sqrt{1 - v^2}$ = Lorentz Factor

Eq. 7: Kinetic Energy

$$KE = \frac{1}{2}mv^2$$

KE = kinetic energy measured in Joules

m = mass

v = velocity

Eq. 8: Einstein's Mass-Energy Equation

$$E = mc^2$$

E = energy

m = mass

c = speed of light

Eq. 9: Einstein's Mass-Energy Equation

$E = mc^2$

E = energy

m = mass

c = speed of light

Eq. 10: Dirac Equation

$$(\beta mc^2 + c \sum_{n=1}^{3} \alpha_n p_n)\psi(x,t) = \frac{i h \partial \psi(x,t)}{2\pi \partial t}$$

In this equation,

ψ = wave function for the electron with spacetime coordinates x, t

p = momentum

c = speed of light

h = Planck's Constant, $6.62607015 \times 10^{-34}$ joule – seconds

Eq. 11: Minkowski's timelike spacetime formula

$$(Spacetime\ Interval)^2 = (Time\ Interval)^2 - (Space\ Interval)^2$$

$$\Delta S^2 = \Delta t^2 - \Delta x^2$$

ΔS = spacetime interval

Δt = time interval between two events

Δx = space interval

Eq. 12: Newton's Law of Gravitation

$$F = G \frac{m_1 m_2}{r^2}$$

F = gravitational force between two objects

G = gravitational constant, $G = 6.6743 \times 10^{-11} m^3 kg^{-1} s^{-2}$

r = distance between the objects

$m_1 m_2$ = masses of separate objects

Eq. 13: Einstein's General Relativity Equation

$$G_{ab} = KT_{ab}$$

G_{ab} = ten-component Curvature Tensor representing space-time curvature

K = constant equal to $\frac{8\pi G}{c^4}$, where G is the gravitational constant

T_{ab} = ten-component Momenergy Tensor representing mass-energy and energy-density

Eq. 14: Hubble's Law

$$v = H_0 d$$

v = receding velocity of a star or galaxy

H_0 = Hubble's constant, 70 km/s/Mpc (where 1 Mpc = 106 parsec = 3.26 × 106 lightyears)

d = estimated distance from Earth

REFERENCES

Bahcall, N. (2015). *Hubble's Law and the Expanding Universe*. Proceedings of the National Academy of Sciences. www.pnas.org/doi/10.1073/pnas.1424299112

Belendez, A. (2015). *Faraday and the Electromagnetic Theory of Light*. Open Mind BBVA. https://www.bbvaopenmind.com/en/science/leading-figures/faraday-electromagnetic-theory-light/

Briggs, A. (2020). *What is Dark Energy?* Earthsky. https://earthsky.org/space/

Briggs, A. (2020). *What Is Dark Matter?* Earthsky. https://earthsky.org/astronomy-essentials/definition-what-is-dark-matter/

Clavin, W. (2020). *Where is Dark Matter Hiding?* Caltech. https://magazine.caltech.edu/post/where-is-dark-matter-hiding

Egdall, Ira (2014). *Einstein Relatively Simple: Our Universe Revealed in Everyday Language*. World Scientific.

Einstein, Albert (2016). *Relativity: The Special and the General Theory*. Digireads Publishing.

Gerritsma, R, Kirchmair, G, Zahringer, F, Solano, E, Blatt R, Roos, F. (2009, September 3). Quantum Simulation of Dirac Equation. Retrieved from https://arxiv.org/abs/0909.0674

Groves, L. R. (1975). *Now It Can Be Told: The Story of The Manhattan Project*. World Scientific. https://www.worldscientific.com/worldscibooks/10.1142/5654

Inglis-Arkell, E. (2013). *The 200-Year-Old Mystery of Mercury's Orbit — Solved*. Gizmodo.com. https://gizmodo.com/the-200-year-old-mystery-of-mercurys-orbit-solved-1458642219

Kelly, C. (2005). *Remembering the Manhattan Project*. World Scientific. https://www.worldscienti!c.com/worldscibooks/10.1142/5654

Lasky, R. (2003). *How Does Relativity Resolve the Twin Paradox?* Scientific American. https://www.scientificamerican.com/article/how-does-relativity-theor/

Learn, J. (2021). *Schrödinger's Cat Experiment and the Conundrum That Rules Modern Physics*. Discover Magazine. https://www.discovermagazine.com/the-sciences/schroedingers-cat-experiment-and-the-conundrum-that-rules-modern-physics

Lincoln, D. (2022). *Killing AEther: The Michelson-Morley Experiment*. Wondrium Daily. https://www.wondriumdaily.com/killing-aEther-the-michelson-morley-experiment/

Mann, A. (2020). *What is Dark Matter?* Live Science. https://www.livescience.com/dark-matter.htm

Metcalfe, T. (2023). *What Was the Manhattan Project?* Scientific American. https://www.scientificamerican.com/article/what-was-the-manhattan-project/

O'Callaghan, J. (2021). *What Is the Electromagnetic Spectrum?* Space.com. https://www.space.com/what-is-the-electromagnetic-spectrum

O'Connor, J.J. (2004). *Newton's Bucket*. Spacetimecenter.org. http://spacetimecentre.org/vpetkov/courses/Newton_bucket.html

Pössel, M. (2005). *The Elevator, the Rocket, and Gravity: The Equivalence Principle*. Einstein Online. www.einstein-online.info/en/spotlight/equivalence_principle/

Pratt, Carl J. (2021). *Quantum Physics for Beginners*. Ippoceronte Publishing Project. Da Capo Press.

Redd, N. T. (2017). *What Is Wormhole Theory?* Space.com. https://www.space.com/20881-wormholes.html

Siegel, E. (2021). *How To Understand Einstein's Equation for General Relativity*. Big Think. https://bigthink.com/starts-with-a-bang/einstein-general-theory-relativity-equation/

Stein, V. (2023). *What Is String Theory?* Space.com. https://www.space.com/17594-string-theory.html

Szyk, B. (2023). *Time Dilation Calculator*. Omnicalculator.com. https://www.omnicalculator.com/physics/time-dilation

Tillman, N. (2013). *What is Dark Energy?* Space.com. https://www.space.com/20929-dark-energy.htm

Tillman, N. (2023). *Black Holes: Everything You Need to Know*. Space.com.

https://www.space.com/15421-black-holes-facts-formation-discovery-sdcmp.html

Urone, P. (2000). *Length Contraction*. Physics Libre Texts. https://phys.libretexts.org/Bookshelves/College_Physics/College_Physics_1e_(OpenStax)/28%3A_Special_Relativity/28.03%3A_Length_Contraction

Webb, R. (2008). *A Relative Success*. Nature. https://www.nature.com/articles/milespin04

Wells, S. (2021). *Dark Energy: The Physics – Breaking Force that Shapes Our Universe, Explained*. Inverse. https://www.inverse.com/science/darkmatter-energy-explained

Wright, A. (2013). *Across The Universe*. Nature Physics. https://www.nature.com/articles/nphys2629

IMAGE CREDITS

Fig. 1: "Milky Way", by Pablo Carlos Budassi, licensed under CC By 4.0. Text added to original.

Fig. 2: "Electromagnetic radiation spectrum with wavelength, frequency, and energy", by Philip Ronan, licensed under CC By 4.0.

Fig. 3: Author's work.

Fig. 4: "German born theoretical physicist Albert Einstein", by Lucien Chavan, licensed under CC By 4.0.

Fig. 5: "Function of a light clock", by Michael Schmid, licensed under CC By 4.0.

Fig. 6: "Train icon for multiple unit", by Alancrh, licensed under CC By 4.0. Lightning and stick figures added to the original.

Fig. 7: "Schrodinger's cat thought experiment", by Dhatfield, licensed under CC By 4.0.

Fig. 8: "Some Simple World Lines", by Tom Weideman, licensed under CC By 4.0.

Fig. 9: "Mercury's Wobble", by University of Florida Astronomy Department, licensed under CC By 4.0.

Fig. 10: Author's work.

Fig. 11: "Albert Einstein and his first wife, Mileva", author unknown, licensed under CC By 4.0.

Fig. 12: "Thought experiment in Einstein's elevator with a ray of light", by Mathieu Rouaud, licensed under CC By 4.0.

Fig. 13: "Ehrenfest paradox", by Prokaryotic Caspase Homolog, licensed under CC By 4.0.

Fig. 14: "Latitude and Longitude of the Earth", by Djexplo, licensed under CC By 4.0.

Fig. 15: Author's work.

Fig. 16: "Spacetime lattice analogy", by Mysid, licensed under CC By 4.0.

Fig. 17: "Light rays from a distant star", by Hanoch Gutfreund, licensed under CC By 4.0.

Fig. 18: "Newton's water bucket experiment", by Conrad Ranzan, licensed under CC By 4.0.

Fig. 19: "EHT Sagittarius A black hole", by EHT Collaboration, licensed under CC By 4.0.

Fig. 20: "Spitzer Captures Messier 87", by NASA, JPL-Caltech/IPAC, licensed under CC By 4.0.

Fig. 21: "Standard Model of Particle Physics", by Cush, licensed under CC By 4.0.

Made in United States
Troutdale, OR
05/05/2024